Health and Safety
Commission

Workplace (Health, Safety and Welfare) Regulations 1992

Approved Code of Practice and Guidance
L24

London: HMSO

Enquiries regarding this or any other HSE
publications should be made to the HSE
Information Centre at the following address:

HSE Information Centre
Broad Lane
Sheffield
S3 7HQ
Tel: (0742) 892345
Fax: (0742) 892333

Contents

Schedules to the Regulations

Appendices to the guidance

By virtue of Section 16(1) of the Health and Safety at Work etc Act 1974 ("the 1974 Act") and with the consent of the Secretary of State for Employment, the Health and Safety Commission has on 7 December 1992 approved a Code of Practice which provides practical guidance with respect to the provisions of the Workplace (Health, Safety and Welfare) Regulations. The Code of Practice consists of those paragraphs which are identified as such in the document entitled *Workplace health, safety and welfare*.

The Code of Practice comes into effect on 1 January 1993.

Signed

T A GATES
Secretary to the Health and Safety Commission

8 December 1992

This document contains the Workplace (Health, Safety and Welfare) Regulations 1992, together with an Approved Code of Practice and additional guidance.

The Approved Code of Practice consists of the paragraphs which are identified as such in the document entitled *Workplace health, safety and welfare*. It has been approved by the Health and Safety Commission, with the consent of the Secretary of State, under section 16 of the Health and Safety at Work etc Act 1974 for the purpose of providing practical guidance with respect to the Workplace (Health, Safety and Welfare) Regulations 1992. It has been drawn up on the basis of consultations between representatives of the Confederation of British Industry, the Trades Union Congress, local authorities, other interested parties, and the Health and Safety Executive.

Although failure to comply with any provision of the Approved Code of Practice is not in itself an offence, the failure may be taken to Court in criminal proceedings as proof that a person has contravened the regulation to which the provision relates. In such a case, however, it will be open to that person to satisfy a Court that he or she has complied with the regulation in some other way.

Words and expressions which are defined in the Health and Safety at Work etc Act 1974 or in the Workplace (Health, Safety and Welfare) Regulations 1992 have the same meaning in the Approved Code of Practice unless the context requires otherwise.

Guidance which does not form part of the Approved Code of Practice is identified as such in this document. Where the guidance refers to another publication, a number in small print indicates that further details of the publication are given in Appendix 1.

In addition to the Workplace Regulations, the following Regulations (which also stem from recent European Directives) are also due to come into force in 1993:

The Management of Health and Safety at Work Regulations 1992[2-3]

The Provision and Use of Work Equipment Regulations 1992[11-12]

The Personal Protective Equipment at Work Regulations 1992[13-14]

The Health and Safety (Display Screen Equipment) Regulations 1992[39-40]

The Manual Handling Operations Regulations 1992

All of these except the last are referred to in this document; details are given in Appendix 1. A guide to the Manual Handling Operations Regulations 1992 is also available from HMSO (*Manual Handling* ISBN 0 11 886335 5).

Citation and commencement

(1)   These Regulations may be cited as the Workplace (Health, Safety and Welfare) Regulations 1992.

(2)   Subject to paragraph (3), these Regulations shall come into force on 1 January 1993.

(3)   Regulations 5 to 27 and the Schedules shall come into force on 1 January 1996 with respect to any workplace or part of a workplace which is not -

(a)   a new workplace; or

(b)   a modification, an extension or a conversion.

Guidance

1   The Regulations come into effect in two stages. Workplaces which are used for the first time after 31 December 1992, and modifications, extensions and conversions started after that date, should comply as soon as they are in use. In existing workplaces (apart from any modifications) the Regulations take effect on 1 January 1996 and the laws in Schedule 2 will continue to apply until that date.

Interpretation

(1)   In these Regulations, unless the context otherwise requires -

"new workplace" means a workplace used for the first time as a workplace after 31 December 1992;

"public road" means (in England and Wales) a highway maintainable at public expense within the meaning of section 329 of the Highways Act 1980 and (in Scotland) a public road within the meaning assigned to that term by section 151 of the Roads (Scotland) Act 1984;

"traffic route" means a route for pedestrian traffic, vehicles or both and includes any stairs, staircase, fixed ladder, doorway, gateway, loading bay or ramp;

"workplace" means, subject to paragraph (2), any premises or part of premises which are not domestic premises and are made available to any person as a place of work, and includes -

(a)   any place within the premises to which such person has access while at work; and

(b)   any room, lobby, corridor, staircase, road or other place used as a means of access to or egress from that place of work or where facilities are provided for use in connection with the place of work other than a public road;

but shall not include a modification, an extension or a conversion of any of the above until such modification, extension or conversion is completed.

(2)   Any reference in these Regulations, except in paragraph (1), to a modification, an extension or a conversion is a reference, as the case may be, to a modification, an extension or a conversion of a workplace started after 31 December 1992.

(3)   Any requirement that anything done or provided in pursuance of these Regulations shall be suitable shall be construed to include a requirement that it is suitable for any person in respect of whom such thing is so done or provided.

(4)   Any reference in these Regulations to -

(a)   a numbered regulation or Schedule is a reference to the regulation in or Schedule to these Regulations so numbered; and

(b)   a numbered paragraph is a reference to the paragraph so numbered in the regulation in which the reference appears.

2   These Regulations apply to a very wide range of workplaces, not only factories, shops and offices but, for example, schools, hospitals, hotels and places of entertainment. The term workplace also includes the common parts of shared buildings, private roads and paths on industrial estates and business parks, and temporary work sites (but not construction sites).

3   'Workplace' is defined in regulation 2(1). Certain words in the definition are themselves defined in sections 52 and 53 of the Health and Safety at Work etc Act 1974. In brief:

'Work' means work as an employee or self-employed person, and also:

(a)   work experience on certain training schemes (Health and Safety (Training for Employment) Regulations 1990 No 138 regulation 3);

(b)   training which includes operations involving ionising radiations (Ionising Radiations Regulations 1985 No 1333 regulation 2(1));

(c)   activities involving genetic manipulation (Genetic Manipulation Regulations 1989 No 1810 regulation 3); and

(d)   work involving the keeping and handling of a listed pathogen (Health and Safety (Dangerous Pathogens) Regulations 1981, S.I. 1981 No.1011 regulation 9).

'Premises' means any place (including an outdoor place).

'Domestic Premises' means a private dwelling. These Regulations do not apply to domestic premises, and do not therefore cover homeworkers. They do, however, apply to hotels, nursing homes and the like, and to parts of workplaces where 'domestic' staff are employed such as the kitchens of hostels or sheltered accommodation.

4   These Regulations aim to ensure that workplaces meet the health, safety and welfare needs of each member of the workforce which may include people with disabilities. Several of the Regulations require things to be 'suitable' as defined in regulation 2(3) in a way which makes it clear that traffic routes, facilities and workstations which are used by people with disabilities should be suitable for them to use.

5   Building Regulations contain requirements which are intended to make new buildings accessible to people with limited mobility, or impaired sight or hearing. There is also a British Standard on access to buildings for people with disabilities[1].

New workplaces

6   A 'new workplace' is one that is taken into use for the **first time** after 31 December 1992. Therefore if a building was a workplace at any time in the past it is not a new workplace (although it may be a conversion).

Modifications, extensions and conversions

7   Any modification or extension started after 31 December 1992 should comply with any relevant requirements of these Regulations as soon as it is in use. This applies only to the actual modification or extension. The rest of the workplace should comply as from 1 January 1996 and until that date the laws listed in Schedule 2 will continue to apply. A 'modification' includes any alteration but not a simple replacement.

8   The whole of any conversion started after 31 December 1992 should comply as soon as it is in use. 'Conversion' is not defined and is therefore any workplace which would ordinarily be considered to be a conversion. Examples of conversions include:

(a)   a large building converted into smaller industrial units. Each unit is a 'conversion';

(b)   a private house, or part of a house, converted into a workplace;

(c)   workplaces which undergo a radical change of use involving structural alterations.

Note: certain modifications, extensions and conversions will also be subject to Building Regulations and may need planning consent. Advice can be obtained from the local authority.

Application of these Regulations

(1)   These Regulations apply to every workplace but shall not apply to -

(a)   a workplace which is or is in or on a ship within the meaning assigned to that word by regulation 2(1) of the Docks Regulations 1988;

(b)   a workplace where the only activities being undertaken are building operations or works of engineering construction within, in either case, section 176 of the Factories Act 1961 and activities for the purpose of or in connection with the first-mentioned activities;

(c)   a workplace where the only activities being undertaken are the exploration for or extraction of mineral resources; or

(d)   a workplace which is situated in the immediate vicinity of another workplace or intended workplace where exploration for or extraction of mineral resources is being or will be undertaken, and where the only activities being undertaken are activities preparatory to, for the purposes of, or in connection with such exploration for or extraction of mineral resources at that other workplace.

(2)   In their application to temporary work sites, any requirement to ensure a workplace complies with any of regulations 20 to 25 shall have effect as a requirement to so ensure so far as is reasonably practicable.

(3)   As respects any workplace which is or is in or on an aircraft, locomotive or rolling stock, trailer or semi-trailer used as a means of transport or a vehicle for which a licence is in force under the Vehicles (Excise) Act 1971 or a vehicle exempted from duty under that Act -

(a)   regulations 5 to 12 and 14 to 25 shall not apply to any such workplace; and

(b)   regulation 13 shall apply to any such workplace only when the aircraft, locomotive or rolling stock, trailer or semi-trailer or vehicle is stationary inside a workplace and, in the case of a vehicle for which a licence is in force under the Vehicles (Excise) Act 1971, is not on a public road.

(4)   As respects any workplace which is in fields, woods or other land forming part of an agricultural or forestry undertaking but which is not inside a building and is situated away from the undertaking's main buildings -

(a)   regulations 5 to 19 and 23 to 25 shall not apply to any such workplace; and

(b)   any requirement to ensure that any such workplace complies with any of regulations 20 to 22 shall have effect as a requirement to so ensure so far as is reasonably practicable.

Means of transport

9   All operational ships, boats, hovercraft, aircraft, trains and road vehicles are excluded from these Regulations, except that regulation 13 applies to aircraft, trains and road vehicles when stationary in a workplace (but not when on a public road). Non-operational means of transport used as, for example, restaurants or tourist attractions, are subject to these Regulations.

Extractive industries (mines, quarries etc)

10   These Regulations do not apply to mines, quarries or other mineral extraction sites, including those off-shore. Nor do they apply to any related workplace on the same site. Other legislation applies to this sector.

Construction sites

11   Construction sites (including site offices) are excluded from these Regulations. Where construction work is in progress within a workplace, it can be treated as a construction site and so excluded from these Regulations, if it is fenced off; otherwise, these Regulations and Construction Regulations will both apply.

Temporary work sites

12   At temporary work sites the requirements of these Regulations for sanitary conveniences, washing facilities, drinking water, clothing accommodation, changing facilities and facilities for rest and eating meals (regulations 20-25) apply 'so far as is reasonably practicable'. Temporary work sites include:

(a)   work sites used only infrequently or for short periods; and

(b)   fairs and other structures which occupy a site for a short period.

Farming and forestry

13   Agricultural or forestry workplaces which are outdoors and away from the undertaking's main buildings are excluded from these Regulations, except for the requirements on sanitary conveniences, washing facilities and drinking water (regulations 20-22) which apply 'so far as is reasonably practicable'.

4

Requirements under these Regulations

(1)   Every employer shall ensure that every workplace, modification, extension or conversion which is under his control and where any of his employees work complies with any requirement of these Regulations which -

(a)   applies to that workplace or, as the case may be, to the workplace which contains that modification, extension or conversion; and

(b)   is in force in respect of the workplace, modification, extension or conversion.

(2)   Subject to paragraph (4), every person who has, to any extent, control of a workplace, modification, extension or conversion shall ensure that such workplace, modification, extension or conversion complies with any requirement of these Regulations which -

(a)   applies to that workplace or, as the case may be, to the workplace which contains that modification, extension or conversion;

(b)   is in force in respect of the workplace, modification, extension, or conversion; and

(c)   relates to matters within that person's control.

(3)   Any reference in this regulation to a person having control of any workplace, modification, extension or conversion is a reference to a person having control of the workplace, modification, extension or conversion in connection with the carrying on by him of a trade, business or other undertaking (whether for profit or not).

(4)   Paragraph (2) shall not impose any requirement upon a self-employed person in respect of his own work or the work of any partner of his in the undertaking.

(5)   Every person who is deemed to be the occupier of a factory by virtue of section 175(5) of the Factories Act 1961 shall ensure that the premises which are so deemed to be a factory comply with these Regulations.

14   Employers have a general duty under Section 2 of the Health and Safety at Work etc Act 1974 to ensure, so far as is reasonably practicable, the health, safety and welfare of their employees at work. Persons in control of non-domestic premises also have a duty under Section 4 of the Act towards people who are not their employees but use their premises. (These Sections are reproduced in Appendix 2.) These Regulations expand on these duties. They are intended to protect the health and safety of everyone in the workplace, and to ensure that adequate welfare facilities are provided for people at work.

15   Employers have a duty to ensure that workplaces under their control comply with these Regulations. Tenant employers are responsible for ensuring that the workplace which they control complies with the Regulations, and that the facilities required by the Regulations are provided, for example that sanitary conveniences are sufficient and suitable, adequately ventilated and lit and kept in a clean and orderly condition. Facilities should be readily accessible but it is not essential that they are within the employer's own workplace; arrangements can be made to use facilities provided by, for example, a landlord or a neighbouring business but the employer is responsible for ensuring that they comply with the Regulations.

16   People other than employers also have duties under these Regulations if they have control, to any extent, of a workplace. For example, owners and landlords (of business premises) should ensure that common parts, common facilities, common services and means of access within their control, comply with the Regulations. Their duties are limited to matters which are within their control. For example, an owner who is responsible for the general condition of a lobby, staircase and landings, for shared toilets provided for tenants' use, and for maintaining ventilation plant, should ensure that those parts and plant comply with these Regulations. However, the owner is not responsible under these Regulations for matters outside his control, for example a spillage caused by a tenant or shortcomings in the day-to-day cleaning of sanitary facilities where this is the tenants' responsibility. Tenants should cooperate with each other, and with the landlord, to the extent necessary to ensure that the requirements of the Regulations are fully met.

17   In some cases, measures additional to those indicated in the Regulations and the Approved Code of Practice may be necessary in order to fully comply with general duties under the Health and Safety at Work etc Act. The Management of Health and Safety at Work Regulations 1992[2-3] require employers and self-employed people to assess risks; an associated Approved Code of Practice states that it is always best if possible to avoid a risk altogether, and that work should, where possible, be adapted to the individual. A risk assessment may show that the workplace or the work should be reorganised so that the need for people to work, for example, at an unguarded edge or to work in temperatures which may induce stress does not arise in the first place.

18   It is often useful to seek the views of workers before and after changes are introduced, for example on the design of workstations, the choice of work chairs, and traffic management systems such as one-way vehicle routes or traffic lights. As well as promoting good relations, consultation can result in better decisions and in some cases help employers avoid making expensive mistakes. The Management of Health and Safety at Work Regulations[2-3] extend the law which requires employers to consult employees' safety representatives on matters affecting health and safety.

19   Where employees work at a workplace which is not under their employer's control, their employer has no duty under these Regulations, but should (as part of his or her general duties under the Health and Safety at Work etc Act 1974) take any steps necessary to ensure that sanitary conveniences and washing facilities will be available. It may be necessary to make arrangements for the use of facilities already provided on site, or to provide temporary facilities. This applies, for example, to those who employ seasonal agricultural workers to work on someone else's land.

Maintenance of workplace, and of equipment, devices and systems

(1)   The workplace and the equipment, devices and systems to which this regulation applies shall be maintained (including cleaned as appropriate) in an efficient state, in efficient working order and in good repair.

(2)   Where appropriate, the equipment, devices and systems to which this regulation applies shall be subject to a suitable system of maintenance.

(3)   The equipment, devices and systems to which this regulation applies are -

(a)   equipment and devices a fault in which is liable to result in a failure to comply with any of these Regulations; and

(b)   mechanical ventilation systems provided pursuant to regulation 6 (whether or not they include equipment or devices within sub-paragraph (a) of this paragraph).

20   The workplace, and the equipment and devices mentioned in these Regulations, should be maintained in an efficient state, in efficient working order and in good repair. 'Efficient' in this context means efficient from the view of health, safety and welfare (not productivity or economy). If a potentially dangerous defect is discovered, the defect should be rectified immediately or steps should be taken to protect anyone who might be put at risk, for example by preventing access until the work can be carried out or the equipment replaced. Where the defect does not pose a danger but makes the equipment unsuitable for use, for example a sanitary convenience with a defective flushing mechanism, it may be taken out of service until it is repaired or replaced, but if this would result in the number of facilities being less than that required by the Regulations the defect should be rectified without delay.

21   Steps should be taken to ensure that repair and maintenance work is carried out properly.

22   Regulation 5(2) requires a system of maintenance where appropriate, for certain equipment and devices and for ventilation systems. A suitable system of maintenance involves ensuring that:

(a)   regular maintenance (including, as necessary, inspection, testing, adjustment, lubrication and cleaning) is carried out at suitable intervals;

(b)   any potentially dangerous defects are remedied, and that access to defective equipment is prevented in the meantime;

(c)   regular maintenance and remedial work is carried out properly; and

(d)   a suitable record is kept to ensure that the system is properly implemented and to assist in validating maintenance programmes.

23   Examples of equipment and devices which require a system of maintenance include emergency lighting, fencing, fixed equipment used for window cleaning, anchorage points for safety harnesses, devices to limit the opening of windows, powered doors, escalators and moving walkways.

24   The frequency of regular maintenance, and precisely what it involves, will depend on the equipment or device concerned. The likelihood of defects developing, and the foreseeable consequences, are highly relevant. The age and condition of equipment, how it is used and how often it is used should also be taken into account. Sources of advice include published HSE guidance, British and EC standards and other authoritative guidance, manufacturers' information and instructions, and trade literature.

25   The Management of Health and Safety at Work Regulations 1992 include requirements on the competence of people whom employers appoint to assist them in matters affecting health and safety and on employees' duties to report serious dangers and shortcomings in health and safety precautions[2-3].

26   There are separate HSE publications covering maintenance of escalators and window access equipment[4-7].

27   Advice on systems of maintenance for buildings can be found in a British Standard[8] and in publications by the Chartered Institution of Building Services Engineers (CIBSE)[9-10]. The maintenance of work equipment and personal protective equipment is addressed in other Regulations[11-14].

Ventilation

(1)   Effective and suitable provision shall be made to ensure that every enclosed workplace is ventilated by a sufficient quantity of fresh or purified air.

(2)   Any plant used for the purpose of complying with paragraph (1) shall include an effective device to give visible or audible warning of any failure of the plant where necessary for reasons of health or safety.

(3)   This regulation shall not apply to any enclosed workplace or part of a workplace which is subject to the provisions of -

(a)   section 30 of the Factories Act 1961;

(b)   regulations 49 to 52 of the Shipbuilding and Ship-Repairing Regulations 1960;

(c)   regulation 21 of the Construction (General Provisions) Regulations 1961;

(d)   regulation 18 of the Docks Regulations 1988.

28   Enclosed workplaces should be sufficiently well ventilated so that stale air, and air which is hot or humid because of the processes or equipment in the workplace, is replaced at a reasonable rate.

29   The air which is introduced should, as far as possible, be free of any impurity which is likely to be offensive or cause ill health. Air which is taken from the outside can normally be considered to be 'fresh', but air inlets for ventilation systems should not be sited where they may draw in excessively contaminated air (for example close to a flue, an exhaust ventilation system outlet, or an area in which vehicles manoeuvre). Where necessary the inlet air should be filtered to remove particulates.

30   In many cases, windows or other openings will provide sufficient ventilation in some or all parts of the workplace. Where necessary, mechanical ventilation systems should be provided for parts or all of the workplace, as appropriate.

31   Workers should not be subject to uncomfortable draughts. In the case of mechanical ventilation systems it may be necessary to control the direction or velocity of air flow. Workstations should be re-sited or screened if necessary.

32   In the case of mechanical ventilation systems which recirculate air, including air conditioning systems, recirculated air should be adequately filtered to remove impurities. To avoid air becoming unhealthy, purified air should have some fresh air added to it before being recirculated. Systems should therefore be designed with fresh air inlets which should be kept open.

33   Mechanical ventilation systems (including air conditioning systems) should be regularly and properly cleaned, tested and maintained to ensure that they are kept clean and free from anything which may contaminate the air.

34   The requirement of regulation 6(2) for a device to give warning of breakdowns applies only 'where necessary for reasons of health or safety'. It will not apply in most workplaces. It will, however, apply to 'dilution ventilation' systems used to reduce concentrations of dust or fumes in the atmosphere, and to any other situation where a breakdown in the ventilation system would be likely to result in harm to workers.

35   Regulation 6 covers general workplace ventilation, not local exhaust ventilation for controlling employees' exposure to asbestos, lead, ionising radiations or other substances hazardous to health. There are other health and safety regulations and approved codes of practice on the control of such substances[15-22].

36   It may not always be possible to remove smells coming in from outside, but reasonable steps should be taken to minimise them. Where livestock is kept, smells may be unavoidable, but they should be controlled by good ventilation and regular cleaning.

37   Where a close, humid atmosphere is necessary, for example in mushroom growing, workers should be allowed adequate breaks in a well-ventilated place.

38   The fresh air supply rate should not normally fall below 5 to 8 litres per second, per occupant. Factors to be considered include the floor area per person, the processes and equipment involved, and whether the work is strenuous.

39   More detailed guidance on ventilation is contained in HSE publications[23-24] and in publications by the Chartered Institution of Building Services Engineers[25-27].

40   Guidance on the measures necessary to avoid legionnaires' disease, caused by bacteria which can grow in water cooling towers and elsewhere, is covered in separate HSE publications[28-29] and in a CIBSE publication[30].

41   The legislation referred to in regulation 6(3) deals with what are known as 'confined spaces' where breathing apparatus may be necessary.

Temperature in indoor workplaces

(1)   During working hours, the temperature in all workplaces inside buildings shall be reasonable.

(2)   A method of heating or cooling shall not be used which results in the escape into a workplace of fumes, gas or vapour of such character and to such extent that they are likely to be injurious or offensive to any person.

(3)   A sufficient number of thermometers shall be provided to enable persons at work to determine the temperature in any workplace inside a building.

42   The temperature in workrooms should provide reasonable comfort without the need for special clothing. Where such a temperature is impractical because of hot or cold processes, all reasonable steps should be taken to achieve a temperature which is as close as possible to comfortable. 'Workroom' in paragraphs 43 to 49 means a room where people normally work for more than short periods.

43   The temperature in workrooms should normally be at least 16 degrees Celsius unless much of the work involves severe physical effort in which case the temperature should be at least 13 degrees Celsius. These temperatures may not, however, ensure reasonable comfort, depending on other factors such as air movement and relative humidity. These temperatures refer to readings taken using an ordinary dry bulb thermometer, close to workstations, at working height and away from windows.

44   Paragraph 43 does not apply to rooms or parts of rooms where it would be impractical to maintain those temperatures, for example in rooms which have to be open to the outside, or where food or other products have to be kept cold. In such cases the temperature should be as close to those mentioned in paragraph 43 as is practical. In rooms where food or other products have to be kept at low temperatures this will involve such measures as:

(a)   enclosing or insulating the product

(b)   pre-chilling the product

(c)   keeping chilled areas as small as possible

(d)   exposing the product to workroom temperatures as briefly as possible.

45   Paragraphs 43 and 44 do not apply to rooms where lower maximum room temperatures are required in other laws, such as the Fresh Meat Export (Hygiene and Inspection) Regulations 1987. It should be noted however that general Food Hygiene Regulations do not specify maximum room temperatures.

46   Where the temperature in a workroom would otherwise be uncomfortably high, for example because of hot processes or the design of the building, all reasonable steps should be taken to achieve a reasonably comfortable temperature, for example by:

(a)   insulating hot plants or pipes;

(b)   providing air cooling plant;

(c)   shading windows;

(d)   siting workstations away from places subject to radiant heat.

47   Where a reasonably comfortable temperature cannot be achieved throughout a workroom, local heating or cooling (as appropriate) should be provided. In extremely hot weather fans and increased ventilation may be used instead of local cooling. Insulated duckboards or other floor coverings should be provided where workers have to stand for long periods on cold floors unless special footwear is provided which prevents discomfort. Draughts should be excluded and self-closing doors installed where such measures are practical and would reduce discomfort.

48   Where, despite the provision of local heating or cooling, workers are exposed to temperatures which do not give reasonable comfort, suitable protective clothing and rest facilities should be provided. Where practical there should be systems of work (for example, task rotation) to ensure that the length of time for which individual workers are exposed to uncomfortable temperatures is limited.

49   In parts of the workplace other than workrooms, such as sanitary facilities or rest facilities, the temperature should be reasonable in all the circumstances including the length of time people are likely to be there. Changing rooms and shower rooms should not be cold.

50   Where persons are required to work in normally unoccupied rooms such as storerooms, other than for short periods, temporary heating should be provided if necessary to avoid discomfort.

51   More detailed guidance on thermal comfort is expected to be published by HSE in 1993.

52   Care needs to be taken when siting temporary heaters so as to prevent burns from contact with hot surfaces. The Provision and Use of Work Equipment Regulations[11-12] require protection from hot surfaces.

53   The Personal Protective Equipment at Work Regulations 1992[13-14] apply to the protective clothing provided for workers' use.

54   Information about Food Hygiene Regulations can be obtained from the Environmental Health Departments of local authorities.

55   Design data relevant to workplace temperatures are published by the Chartered Institution of Building Services Engineers[25].

Injurious or offensive fumes

56   Fixed heating systems should be installed and maintained in such a way that the products of combustion do not enter the workplace. Any heater which produces heat by combustion should have a sufficient air supply to ensure complete combustion. Care should be taken that portable paraffin and liquefied petroleum gas heaters do not produce fumes which will be harmful or offensive.

Thermometers

57   Thermometers should be available at a convenient distance from every part of the workplace to persons at work to enable temperatures to be measured throughout the workplace, but need not be provided in each workroom.

Lighting

(1)   Every workplace shall have suitable and sufficient lighting.

(2)   The lighting mentioned in paragraph (1) shall, so far as is reasonably practicable, be by natural light.

(3)   Without prejudice to the generality of paragraph (1), suitable and sufficient emergency lighting shall be provided in any room in circumstances in which persons at work are specially exposed to danger in the event of failure of artificial lighting.

58   Lighting should be sufficient to enable people to work, use facilities and move from place to place safely and without experiencing eye-strain. Stairs should be well lit in such a way that shadows are not cast over the main part of the treads. Where necessary, local lighting should be provided at individual

workstations, and at places of particular risk such as pedestrian crossing points on vehicular traffic routes. Outdoor traffic routes used by pedestrians should be adequately lit after dark.

59   Dazzling lights and annoying glare should be avoided. Lights and light fittings should be of a type, and so positioned, that they do not cause a hazard (including electrical, fire, radiation or collision hazards). Light switches should be positioned so that they may be found and used easily and without risk.

60   Lights should not be allowed to become obscured, for example by stacked goods, in such a way that the level of light becomes insufficient. Lights should be replaced, repaired or cleaned, as necessary, before the level of lighting becomes insufficient. Fittings or lights should be replaced immediately if they become dangerous, electrically or otherwise.

61   More detailed guidance is given in a separate HSE publication[31]. There are also a number of publications on lighting by the Chartered Institution of Building Services Engineers[32-36].

62   Requirements on lighting are also contained in the Docks Regulations 1988[37-38], the Provision and Use of Work Equipment Regulations 1992[11-12] and the Health and Safety (Display Screen Equipment) Regulations 1992[39-42]. The electrical safety of lighting installations is subject to the Electricity at Work Regulations 1989[43-44].

Natural lighting

63   Windows and skylights should where possible be cleaned regularly and kept free from unnecessary obstructions to admit maximum daylight. Where this would result in excessive heat or glare at a workstation, however, the workstation should be repositioned or the window or skylight should be shaded.

64   People generally prefer to work in natural rather than artificial light. In both new and existing workplaces workstations should be sited to take advantage of the available natural light. Natural lighting may not be feasible where windows have to be covered for security reasons or where process requirements necessitate particular lighting conditions.

Emergency lighting

65   The normal precautions required by these and other Regulations, for example on the prevention of falls and the fencing of dangerous parts of machinery, mean that workers are not in most cases 'specially exposed' to risk if normal lighting fails. Emergency lighting is not therefore essential in most cases. Emergency lighting should however be provided in workrooms where sudden loss of light would present a serious risk, for example if process plant needs to be shut down under manual control or a potentially hazardous process needs to be made safe, and this cannot be done safely without lighting.

66   Emergency lighting should be powered by a source independent from that of normal lighting. It should be immediately effective in the event of failure of the normal lighting, without need for action by anyone. It should provide sufficient light to enable persons at work to take any action necessary to ensure their, and others', health and safety.

67   Fire precautions legislation may require the lighting of escape routes. Advice can be obtained from local fire authorities.

Cleanliness and waste materials

(1)   Every workplace and the furniture, furnishings and fittings therein shall be kept sufficiently clean.

(2)   The surfaces of the floors, walls and ceilings of all workplaces inside buildings shall be capable of being kept sufficiently clean.

(3)   So far as is reasonably practicable, waste materials shall not be allowed to accumulate in a workplace except in suitable receptacles.

68   The standard of cleanliness required will depend on the use to which the workplace is put. For example, an area in which workers take meals would be expected to be cleaner than a factory floor, and a factory floor would be expected to be cleaner than an animal house. However, regulation 12(3) (avoidance of slipping, tripping and falling hazards) should be complied with in all cases.

69   Floors and indoor traffic routes should be cleaned at least once a week. In factories and other workplaces of a type where dirt and refuse accumulates, any dirt and refuse which is not in suitable receptacles should be removed at least daily. These tasks should be carried out more frequently where necessary to maintain a reasonable standard of cleanliness or to keep workplaces free of pests and decaying matter. This paragraph does not apply to parts of workplaces which are normally visited only for short periods, or to animal houses.

70   Interior walls, ceilings and work surfaces should be cleaned at suitable intervals. Except in parts which are normally visited only for short periods, or where any soiling is likely to be light, ceilings and interior walls should be painted, tiled or otherwise treated so that they can be kept clean, and the surface treatment should be renewed when it can no longer be cleaned properly. This paragraph does not apply to parts of workplaces which cannot be safely reached using a 5-metre ladder.

71   Apart from regular cleaning, cleaning should also be carried out when necessary in order to clear up spillages or to remove unexpected soiling of surfaces. Workplaces should be kept free from offensive waste matter or discharges, for example, leaks from drains or sanitary conveniences.

72   Cleaning should be carried out by an effective and suitable method and without creating, or exposing anyone to, a health or safety risk.

73   Care should be taken that methods of cleaning do not expose anyone to substantial amounts of dust, including flammable or explosive concentrations of dusts, or to health or safety risks arising from the use of cleaning agents. The Control of Substances Hazardous to Health Regulations 1988[15-16] are relevant.

74   Absorbent floors, such as untreated concrete or timber, which are likely to be contaminated by oil or other substances which are difficult to remove, should preferably be sealed or coated, for example with a suitable non-slip floor paint. Carpet should also be avoided in such situations.

75   Washable surfaces, and high standards of cleanliness, may be essential for the purposes of infection control (as in the case of post-mortem rooms and pathology laboratories), for the control of exposure to substances hazardous to health or for the purposes of hygiene in the processing or handling of food. In such cases, steps should be taken to eliminate traps for dirt or germs by, for example, sealing joints between surfaces and fitting curved strips or coving along joins between walls and floors and between walls and work surfaces. Further information about food hygiene can be obtained from Environmental Health Departments of local authorities.

Room dimensions and space

(1)   Every room where persons work shall have sufficient floor area, height and unoccupied space for purposes of health, safety and welfare.

(2)   It shall be sufficient compliance with this regulation in a workplace which is not a new workplace, a modification, an extension or a conversion and which, immediately before this regulation came into force in respect of it, was subject to the provisions of the Factories Act 1961, if the workplace does not contravene the provisions of Part I of Schedule 1.

Minimum space

76   Workrooms should have enough free space to allow people to get to and from workstations and to move within the room, with ease. The number of people who may work in any particular room at any one time will depend not only on the size of the room, but on the space taken up by furniture, fittings, equipment, and on the layout of the room. Workrooms, except those where people only work for short periods, should be of sufficient height (from floor to ceiling) over most of the room to enable safe access to workstations. In older buildings with obstructions such as low beams the obstruction should be clearly marked.

77   The total volume of the room, when empty, divided by the number of people normally working in it should be at least 11 cubic metres. In making this calculation a room or part of a room which is more than 3.0 m high should be counted as 3.0 m high. The figure of 11 cubic metres per person is a minimum and may be insufficient if, for example, much of the room is taken up by furniture etc.

78   The figure of 11 cubic metres referred to in paragraph 77 does not apply to:

(a)   Retail sales kiosks, attendants' shelters, machine control cabs or similar small structures, where space is necessarily limited; or

(b)   Rooms being used for lectures, meetings and similar purposes.

79   In a typical room, where the ceiling is 2.4 m high, a floor area of 4.6 m² (for example 2.0 x 2.3 m) will be needed to provide a space of 11 m³. Where the ceiling is 3.0 m high or higher the minimum floor area will be 3.7 m² (for example 2.0 x 1.85 m). (These floor areas are only for illustrative purposes and are approximate.)

80   The floor space per person indicated in paragraph 77 and 79 will not always give sufficient **unoccupied** space, as required by the Regulation. Rooms may need to be larger, or to have fewer people working in them, than indicated in those paragraphs, depending on such factors as the contents and layout of the room and the nature of the work. Where space is limited careful planning of the workplace is particularly important.

Workstations and seating

(1)   Every workstation shall be so arranged that it is suitable both for any person at work in the workplace who is likely to work at that workstation and for any work of the undertaking which is likely to be done there.

(2)   Without prejudice to the generality of paragraph (1), every workstation outdoors shall be so arranged that -

(a)   so far as is reasonably practicable, it provides protection from adverse weather;

(b)   it enables any person at the workstation to leave it swiftly or, as appropriate, to be assisted in the event of an emergency; and

(c)   it ensures that any person at the workstation is not likely to slip or fall.

(3)   A suitable seat shall be provided for each person at work in the workplace whose work includes operations of a kind that the work (or a substantial part of it) can or must be done sitting.

(4)   A seat shall not be suitable for the purposes of paragraph (3) unless -

(a)   it is suitable for the person for whom it is provided as well as for the operations to be performed; and

(b)   a suitable footrest is also provided where necessary.

81   Workstations should be arranged so that each task can be carried out safely and comfortably. The worker should be at a suitable height in relation to the work surface. Work materials and frequently used equipment or controls should be within easy reach, without undue bending or stretching.

82   Workstations including seating, and access to workstations, should be suitable for any special needs of the individual worker, including workers with disabilities.

83   Each workstation should allow any person who is likely to work there adequate freedom of movement and the ability to stand upright. Spells of work which unavoidably have to be carried out in cramped conditions should be kept as short as possible and there should be sufficient space nearby to relieve discomfort.

84   There should be sufficient clear and unobstructed space at each workstation to enable the work to be done safely. This should allow for the manoeuvring and positioning of materials, for example lengths of timber.

85   Seating provided in accordance with regulation 11(3) should where possible provide adequate support for the lower back, and a footrest should be provided for any worker who cannot comfortably place his or her feet flat on the floor.

86   More detailed guidance on seating is given in an HSE publication[45]. There are other HSE publications on visual display units and ergonomics[41-42,46-47].

87   Static and awkward posture at the workstation, the use of undesirable force and an uncomfortable hand grip, often coupled with continuous repetitive work without sufficient rest and recovery, may lead to chronic injury. Guidance is contained in an HSE publication[48].

88   This Regulation covers all workstations. Workstations where visual display units, process control screens, microfiche readers and similar display units are used are subject to the Health and Safety (Display Screen Equipment) Regulations 1992[39-40].

Condition of floors and traffic routes

(1)   Every floor in a workplace and the surface of every traffic route in a workplace shall be of a construction such that the floor or surface of the traffic route is suitable for the purpose for which it is used.

(2)   Without prejudice to the generality of paragraph (1), the requirements in that paragraph shall include requirements that –

(a)   the floor, or surface of the traffic route, shall have no hole or slope, or be uneven or slippery so as, in each case, to expose any person to a risk to his health or safety; and

(b)   every such floor shall have effective means of drainage where necessary.

(3)   So far as is reasonably practicable, every floor in a workplace and the surface of every traffic route in a workplace shall be kept free from obstructions and from any article or substance which may cause a person to slip, trip or fall.

(4)   In considering whether for the purposes of paragraph (2)(a) a hole or slope exposes any person to a risk to his health or safety –

(a)   no account shall be taken of a hole where adequate measures have been taken to prevent a person falling; and

(b)   account shall be taken of any handrail provided in connection with any slope.

(5)   Suitable and sufficient handrails and, if appropriate, guards shall be provided on all traffic routes which are staircases except in circumstances in which a handrail can not be provided without obstructing the traffic route.

89   Floor and traffic routes should be of sound construction and should have adequate strength and stability taking account of the loads placed on them and the traffic passing over them. Floors should not be overloaded.

90   The surfaces of floors and traffic routes should be free from any hole, slope, or uneven or slippery surface which is likely to:

(a)   cause a person to slip, trip or fall;

(b)   cause a person to drop or lose control of anything being lifted or carried; or

(c)   cause instability or loss of control of vehicles and/or their loads.

16

91   Holes, bumps or uneven areas resulting from damage or wear and tear, which may cause a person to trip or fall, should be made good. Until they can be made good, adequate precautions should be taken against accidents, for example by barriers or conspicuous marking. Temporary holes, for example an area where floor boards have been removed, should be adequately guarded. Account should be taken of people with impaired or no sight. Surfaces with small holes (for example metal gratings) are acceptable provided they are not likely to be a hazard. Deep holes into which people may fall are subject to regulation 13 and the relevant section of this Code.

92   Slopes should not be steeper than necessary. Moderate and steep slopes, and ramps used by people with disabilities, should be provided with a secure handrail where necessary.

93   Surfaces of floors and traffic routes which are likely to get wet or to be subject to spillages should be of a type which does not become unduly slippery. A slip-resistant coating should be applied where necessary. Floors near to machinery which could cause injury if anyone were to fall against it (for example a woodworking or grinding machine) should be slip-resistant and be kept free from slippery substances or loose materials.

94   Where possible, processes and plant which may discharge or leak liquids should be enclosed (for example by bunding), and leaks from taps or discharge points on pipes, drums and tanks should be caught or drained away. Stop valves should be fitted to filling points on tank filling lines. Where work involves carrying or handling liquids or slippery substances, as in food processing and preparation, the workplace and work surfaces should be arranged in such a way as to minimise the likelihood of spillages.

95   Where a leak or spillage occurs and is likely to be a slipping hazard, immediate steps should be taken to fence it off, mop it up, or cover it with absorbent granules.

96   Arrangements should be made to minimise risks from snow and ice. This may involve gritting, snow clearing and closure of some routes, particularly outside stairs, ladders and walkways on roofs.

97   Floors and traffic routes should be kept free of obstructions which may present a hazard or impede access. This is particularly important on or near stairs, steps, escalators and moving walkways, on emergency routes, in or near doorways or gangways, and in any place where an obstruction is likely to cause an accident, for example near a corner or junction. Where a temporary obstruction is unavoidable and is likely to be a hazard, access should be prevented or steps should be taken to warn people or the drivers of vehicles of the obstruction by, for example, the use of hazard cones. Where furniture or equipment is being moved within a workplace, it should if possible be moved in a single operation and should not be left in a place where it is likely to be a hazard. Vehicles should not be parked where they are likely to be a hazard. Materials which fall onto traffic routes should be cleared as soon as possible.

98   Effective drainage should be provided where a floor is liable to get wet to the extent that the wet can be drained off. This is likely to be the case in, for example, laundries, textile manufacture (including dyeing, bleaching and finishing), work on hides and skins, potteries and food processing. Drains and channels should be positioned so as to minimise the area of wet floor, and the floor should slope slightly towards the drain. Where necessary to prevent tripping hazards, drains and channels should have covers which should be as near flush as possible with the floor surface.

99   Every open side of a staircase should be securely fenced. As a minimum the fencing should consist of an upper rail at 900 mm or higher and a lower rail.

100   A secure and substantial handrail should be provided and maintained on at least one side of every staircase, except at points where a handrail would obstruct access or egress, as in the case of steps in a theatre aisle. Handrails should be provided on both sides if there is a particular risk of falling, for example where stairs are heavily used, or are wide, or have narrow treads, or where they are liable to be subject to spillages. Additional handrails should be provided down the centre of particularly wide staircases where necessary.

101   Further guidance on slips, trips and falls, and on the containment of pesticides in storage, is contained in separate HSE publications[49-50].

102   Methods of draining and containing toxic, corrosive or highly flammable liquids should not result in the contamination of drains, sewers, watercourses, or groundwater supplies, or put people or the environment at risk. Maximum concentration levels are specified in the Environmental Protection (Prescribed Processes and Substances) Regulations 1991, and the Surface Waters (Dangerous Substances) (Classification) Regulations 1989 and 1992. Consent for discharges may be required under the Environmental Protection Act 1990, the Water Resources Act 1991 and the Water Industry Act 1991.

103   Consideration should be given to providing slip resistant footwear in workplaces where slipping hazards arise despite the precautions set out in paragraph 93.

104   Building Regulations also have requirements on floors and stairs. Advice may be obtained from the local authority. There is also a British Standard on the construction and maximum loading of floors[51].

105   Steep stairways are classed as fixed ladders and are dealt with under regulation 13.

Falls or falling objects

(1)   So far as is reasonably practicable, suitable and effective measures shall be taken to prevent any event specified in paragraph (3).

(2)   So far as is reasonably practicable, the measures required by paragraph (1) shall be measures other than the provision of personal protective equipment, information, instruction, training or supervision.

(3)   The events specified in this paragraph are:

(a)   any person falling a distance likely to cause personal injury;

(b)   any person being struck by a falling object likely to cause personal injury.

(4)   Any area where there is a risk to health or safety from any event mentioned in paragraph (3) shall be clearly indicated where appropriate.

(5)   So far as is practicable, every tank, pit or structure where there is a risk of a person in the workplace falling into a dangerous substance in the tank, pit or structure, shall be securely covered or fenced.

(6)   Every traffic route over, across or in an uncovered tank, pit or structure such as is mentioned in paragraph (5) shall be securely fenced.

(7)   In this Regulation, "dangerous substance" means -

(a)   any substance likely to scald or burn;

(b)   any poisonous substance;

(c)   any corrosive substance;

(d)   any fume, gas or vapour likely to overcome a person; or

(e)   any granular or free-flowing solid substance, or any viscous substance which, in any case, is of a nature or quantity which is likely to cause danger to any person.

106   The consequences of falling from heights or into dangerous substances are so serious that a high standard of protection is required. Secure fencing should normally be provided to prevent people falling from edges, and the fencing should also be adequate to prevent objects falling onto people. Where fencing cannot be provided or has to be removed temporarily, other measures should be taken to prevent falls. Dangerous substances in tanks, pits or other structures should be securely fenced or covered.

107   The guarding of temporary holes, such as an area where floorboards have been removed, is dealt with in paragraph 91 of this Code.

Provision of fencing or covers

108   Secure fencing should be provided wherever possible at any place where a person might fall 2 metres or more. Secure fencing should also be provided where a person might fall less than 2 metres, where there are factors which increase the likelihood of a fall or the risk of serious injury; for example where a traffic route passes close to an edge, where large numbers of people are present, or where a person might fall onto a sharp or dangerous surface or into the path of a vehicle. Tanks, pits or similar structures may be securely covered instead of being fenced.

109   Fencing should be sufficiently high, and filled in sufficiently, to prevent falls (of people or objects) over or through the fencing. As a minimum, fencing should consist of two guard-rails (a top rail and a lower rail) at suitable heights. In the case of fencing installed after 1 January 1993 (but not repairs or partial replacement) the top of the fencing should be at least 1100 mm above the surface from which a person might fall except in cases where lower fencing has been approved by a local authority under Building Regulations.

110   Fencing should be of adequate strength and stability to restrain any person or object liable to fall on to or against it. Untensioned chains, ropes and other non-rigid materials should not be used.

111   Fencing should be designed to prevent objects falling from the edge including items used for cleaning or maintenance. Where necessary an adequate upstand or toeboard should be provided.

112   Covers should be capable of supporting all loads liable to be imposed upon them, and any traffic which is liable to pass over them. They should be of a type which cannot be readily detached and removed, and should not be capable of being easily displaced.

113 Paragraphs 108 to 111 do not apply to edges on roofs or to places to which there is no general access. Nevertheless, secure, adequate fencing should be provided wherever possible in such cases. Tanks, pits or similar structures containing dangerous substances should always be provided with secure fencing or a secure cover.

114 Additional safeguards may be necessary in places where unauthorised entry is foreseeable. A separate HSE publication gives guidance on safeguards for effluent storage in farms[52].

115 Building Regulations also have requirements on fencing. Advice can be obtained from local authorities. There is a British Standard on the construction of fencing[53].

Temporary removal of fencing or covers

116 When an opening or an edge is being used to transfer goods or materials from one level to another, it should be fenced as far as possible. Secure handholds should be provided where workers have to position themselves at an unfenced opening or edge, such as a teagle opening or similar doorway used for the purpose of hoisting or lowering goods. Where the operation necessarily involves the use of an unguarded edge, as little fencing or rail as possible should be removed, and should be replaced as soon as possible.

117 One method of fencing an opening or edge where articles are raised or lowered by means of a lift truck is to provide a special type of fence or barrier which the worker can raise without having to approach the edge, for example by operating a lever, to give the lift truck access to the edge.

118 Covers should be kept securely in place except when they have to be removed for inspection purposes or in order to gain access. Covers should be replaced as soon as possible.

Fixed ladders

119 Fixed ladders should not be provided in circumstances where it would be practical to install a staircase (see paragraph 162 of this Code). Fixed ladders or other suitable means of access or egress should be provided in pits, tanks and similar structures into which workers need to descend. In this Code a 'fixed ladder' includes a steep stairway (a staircase which a person normally descends facing the treads or rungs).

120 Fixed ladders should be of sound construction, properly maintained and securely fixed. Rungs of a ladder should be horizontal, give adequate foothold and not depend solely upon nails, screws or similar fixings for their support.

121 Unless some other adequate handhold exists, the stiles of the ladder should extend at least 1100 mm above any landing served by the ladder or the highest rung used to step or stand on except that in the case of chimneys the stiles should not project into the gas stream.

122 Fixed ladders installed after 31 December 1992 with a vertical distance of more than 6 m should normally have a landing or other adequate resting place at every 6 m point. Each run should, where possible, be out of line with the last run, to reduce the distance a person might fall. Where it is not possible to provide such landings, for example on a chimney, the ladders should only be used by specially trained and proficient people.

123 Where a ladder passes through a floor, the opening should be as small as possible. The opening should be fenced as far as possible, and a gate should be provided where necessary to prevent falls.

124 Fixed ladders at an angle of less than 15 degrees to the vertical (a pitch of more than 75 degrees) which are more than 2.5 m high should where possible be fitted with suitable safety hoops or permanently fixed fall arrest systems. Hoops should be at intervals of not more than 900 mm measured along the stiles, and should commence at a height of 2.5 m above the base of the ladder. The top hoop should be in line with the top of the fencing on the platform served by the ladder. Where a ladder rises less than 2.5 m, but is elevated so that it is possible to fall a distance of more than 2 m, a single hoop should be provided in line with the top of the fencing. Where the top of a ladder passes through a fenced hole in a floor, a hoop need not be provided at that point.

125 Stairs are much safer than ladders, especially when loads are to be carried. A sloping ladder is generally easier and safer to use than a vertical ladder (see Regulation 17 and paragraph 162 of the Code.)

126 British Standards deal with ladders for permanent access[54-55].

Roof work

127 Slips and trips which may be trivial at ground level may result in fatal accidents when on a roof. It is therefore vital that precautions are taken, even when access is only occasional, for example for maintenance or cleaning.

128 As well as falling from the roof edge, there may be a risk of falling through a fragile material. Care should be taken of old materials which may have become fragile because of corrosion. The risks may be increased by moss, lichen, ice, etc. Surfaces may also be deceptive.

129 Where regular access is needed to roofs (including internal roofs, for example a single storey office within a larger building) suitable permanent access should be provided and there should be fixed physical safeguards to prevent falls from edges and through fragile roofs. Where occasional access is required, other safeguards should be provided, for example crawling boards, temporary access equipment etc.

130 A fragile roof or surface is one which would be liable to fracture if a person's weight were to be applied to it, whether by walking, falling on to it or otherwise. All glazing and asbestos cement or similar sheeting should be treated as being fragile unless there is firm evidence to the contrary. Fragile roofs or surfaces should be clearly identified.

131 Construction Regulations contain specific requirements on roof work. An HSE publication gives more detailed advice on roof work[56]. There is also a British Standard on imposed roof loads[57].

Falls into dangerous substances

132 The tanks, pits and structures mentioned in regulation 13(5) are referred to here as 'vessels' and include sumps, silos, vats, and kiers which persons could fall into. (Kiers are fixed vessels which are used for boiling textile materials in workplaces where the printing, bleaching or dyeing of textile materials or waste is carried out.)

133   Every vessel containing a dangerous substance should be adequately protected to prevent a person from falling into it. Vessels installed after 31 December 1992 should be securely covered, or fenced to a height of at least 1100 mm unless the sides extend to at least 1100 mm above the highest point from which people could fall into them. In the case of existing vessels the height should be at least 915 mm or, in the case of atmospheric or open kiers, 840 mm.

Changes of level

134   Changes of level, such as a step between floors, which are not obvious should be marked to make them conspicuous.

Stacking and racking

135   Materials and objects should be stored and stacked in such a way that they are not likely to fall and cause injury. Racking should be of adequate strength and stability having regard to the loads placed on it and its vulnerability to damage, for example by vehicles.

136   Appropriate precautions in stacking and storage include:

(a)   safe palletisation;

(b)   banding or wrapping to prevent individual articles falling out;

(c)   setting limits for the height of stacks to maintain stability;

(d)   regular inspection of stacks to detect and remedy any unsafe stacks; and

(e)   particular instruction and arrangements for irregularly shaped objects.

137   Further guidance on stacking materials is given in HSE publications[58-59].

Loading or unloading vehicles

138   The need for people to climb on top of vehicles or their loads should be avoided as far as possible. Where it is unavoidable, effective measures should be taken to prevent falls.

139   Where a tanker is loaded from a fixed gantry and access is required on to the top of the tanker, fencing should be provided where possible. The fencing may be collapsible fencing on top of the tanker or may form part of the gantry. In the latter case if varying designs of tankers are loaded the fencing should be adjustable, where necessary. Similar fencing should also be provided wherever people regularly go on top of tankers at a particular location, for example for maintenance.

140   Where loaded lorries have to be sheeted before leaving a workplace, suitable precautions should be taken against falls. Where sheeting is done frequently it should be carried out in designated parts of the workplace which are equipped for safe sheeting. Where reasonably practicable, gantries should be provided which lorries can drive under or alongside, so that the load is sheeted from the gantry without any need to stand on the cargo. In other situations safety lines and harnesses should be provided for people on top of the vehicle.

Measures other than fencing, covers, etc

141   When fencing or covers cannot be provided, or have to be removed, effective measures should be taken to prevent falls. Access should be limited to specified people and others should be kept out by, for example, barriers; in high risk situations suitable formal written permit to work systems should be adopted. A safe system of work should be operated which may include the provision and use of a fall arrest system, or safety lines and harnesses, and secure anchorage points. Safety lines should be short enough to prevent injury should a fall occur and the safety line operate. Adequate information, instruction, training and supervision should be given.

142   People should not be allowed into an area where, despite safeguards, they would be in danger, for example from work going on overhead.

143   Systems which do not require disconnection and reconnection of safety harnesses from safety lines, when at risk of falling, should be used in preference to those that do. Where there is no need to approach the edge the length of the line and the position of the anchorage should be such as to prevent the edge being approached.

144   The provision and use of safety harnesses etc are also subject to the Personal Protective Equipment at Work Regulations 1992[13-14]. There are also relevant British Standards[60-61].

Scaffolding

145   Scaffolding and other equipment used for temporary access may either follow the provisions of this code or the requirements of Construction Regulations.

Other Regulations

146   Other Regulations concerning shipyards, docks and agricultural workplaces also contain specific requirements for preventing injury from falls[37-38,62-63]. Those specific requirements stand. Regulation 13 of these Regulations and relevant parts of this Code will also apply to such premises (subject to regulation 3(4) which partially excludes open farmland). However it is not intended that regulation 13 or this Code should be interpreted as overriding or increasing those specific requirements of other Regulations.

Windows, and transparent or translucent doors, gates and walls

(1)   Every window or other transparent or translucent surface in a wall or partition and every transparent or translucent surface in a door or gate shall, where necessary for reasons of health or safety -

(a)   be of safety material or be protected against breakage of the transparent or translucent material; and

(b)   be appropriately marked or incorporate features so as, in either case, to make it apparent.

147   Transparent or translucent surfaces in doors, gates, walls and partitions should be of a safety material or be adequately protected against breakage in the following cases:

(a) in doors and gates, and door and gate side panels, where any part of the transparent or translucent surface is at shoulder level or below;

(b) in windows, walls and partitions, where any part of the transparent or translucent surface is at waist level or below, except in glasshouses where people there will be likely to be aware of the presence of glazing and avoid contact.

This paragraph does not apply to narrow panes up to 250 mm wide measured between glazing beads.

148 'Safety materials' are:

(a) materials which are inherently robust, such as polycarbonates or glass blocks; or

(b) glass which, if it breaks, breaks safely; or

(c) ordinary annealed glass which meets the thickness criteria in the following table:

Nominal thickness	Maximum size
8 mm	1.10 m x 1.10 m
10 mm	2.25 m x 2.25 m
12 mm	3.00 m x 4.50 mm
15 mm	Any size

149 As an alternative to the use of safety materials, transparent or translucent surfaces may be adequately protected against breakage. This may be achieved by means of a screen or barrier which will prevent a person from coming into contact with the glass if he or she falls against it. If a person going through the glass would fall from a height, the screen or barrier should also be designed to be difficult to climb.

150 A transparent or translucent surface should be marked where necessary to make it apparent. The risk of collision is greatest in large uninterrupted surfaces where the floor is at a similar level on each side, so that people might reasonably think they can walk straight through. If features such as mullions, transoms, rails, door frames, large pull or push handles, or heavy tinting make the surface apparent, marking is not essential. Where it is needed, marking may take any form (for example coloured lines or patterns), provided it is conspicuous and at a conspicuous height.

151 The term 'safety glass' is used in a British Standard[64] which is concerned with the breakage of flat glass or flat plastic sheet. Materials meeting that Standard, for example laminated or toughened glass, will break in a way that does not result in large sharp pieces and will fulfil paragraph 148(b) above. 'Safety materials' as used in these Regulations includes safety glass, but also other materials as described in paragraphs 148(a) and (c) above. There is also a British Standard which contains a code of practice for the glazing for buildings[65].

152 Building Regulations also have similar requirements. Advice may be obtained from local authorities.

Windows, skylights and ventilators

(1) No window, skylight or ventilator which is capable of being opened shall be likely to be opened, closed or adjusted in a manner which exposes any person performing such operation to a risk to his health or safety.

(2) No window, skylight or ventilator shall be in a position when open which is likely to expose any person in the workplace to a risk to his health or safety.

153 It should be possible to reach and operate the control of openable windows, skylights and ventilators in a safe manner. Where necessary, window poles or similar equipment should be kept available, or a stable platform or other safe means of access should be provided. Controls should be so placed that people are not likely to fall through or out of the window. Where there is a danger of falling from a height devices should be provided to prevent the window opening too far.

154 Open windows, skylights or ventilators should not project into an area where persons are likely to collide with them. The bottom edge of opening windows should normally be at least 800 mm above floor level, unless there is a barrier to prevent falls.

155 There is a British Standard on windows and skylights[66].

Ability to clean windows etc safely

(1) All windows and skylights in a workplace shall be of a design or be so constructed that they may be cleaned safely.

(2) In considering whether a window or skylight is of a design or so constructed as to comply with paragraph (1), account may be taken of equipment used in conjunction with the window or skylight or of devices fitted to the building.

156 Suitable provision should be made so that windows and skylights can be cleaned safely if they cannot be cleaned from the ground or other suitable surface.

157 Suitable provision includes:

(a) fitting windows which can be cleaned safely from the inside, for example windows which pivot so that the outer surface is turned inwards;

(b) fitting access equipment such as suspended cradles, or travelling ladders with an attachment for a safety harness;

(c) providing suitable conditions for the future use of mobile access equipment, including ladders up to 9 metres long. Suitable conditions are adequate access for the equipment, and a firm level surface in a safe place on which to stand it. Where a ladder over 6 metres long will be needed, suitable points for tying or fixing the ladder should be provided.

(d) suitable and suitably placed anchorage points for safety harnesses.

158 Further guidance on safe window cleaning and access equipment is given in other HSE publications[6-7]. There is also a relevant British Standard[66].

Organisation etc of traffic routes

(1) Every workplace shall be organised in such a way that pedestrians and vehicles can circulate in a safe manner.

(2) Traffic routes in a workplace shall be suitable for the persons or vehicles using them, sufficient in number, in suitable positions and of sufficient size.

(3) Without prejudice to the generality of paragraph (2), traffic routes shall not satisfy the requirements of that paragraph unless suitable measures are taken to ensure that -

(a) pedestrians or, as the case may be, vehicles may use a traffic route without causing danger to the health or safety of persons at work near it;

(b) there is sufficient separation of any traffic route for vehicles from doors or gates or from traffic routes for pedestrians which lead onto it; and

(c) where vehicles and pedestrians use the same traffic route, there is sufficient separation between them.

(4) All traffic routes shall be suitably indicated where necessary for reasons of health or safety.

(5) Paragraph (2) shall apply so far as is reasonably practicable, to a workplace which is not a new workplace, a modification, an extension or a conversion.

159 This section of the Code applies to both new and existing workplaces. In paragraphs 160, 165 and 171 special provision is made for traffic routes in existence before 1 January 1993. This is because it might, in a few cases, otherwise be difficult for existing routes to comply fully with the Code. These special provisions reflect regulation 17(5) which has the effect of requiring existing traffic routes to comply with Regulation 17(2) and 17(3) only to the extent that it is reasonably practicable. 'Traffic route' is defined in regulation 2 as 'a route for pedestrian traffic, vehicles or both and includes any stairs, staircase, fixed ladder, doorway, gateway, loading bay or ramp'.

160 There should be sufficient traffic routes, of sufficient width and headroom, to allow people on foot or in vehicles to circulate safely and without difficulty. Features which obstruct routes should be avoided. On traffic routes in existence before 1 January 1993, obstructions such as limited headroom are acceptable provided they are indicated by, for example, the use of conspicuous tape. Consideration should be given to the safety of people with impaired or no sight.

161 In some situations people in wheelchairs may be at greater risk than people on foot, and special consideration should be given to their safety. Traffic routes used by people in wheelchairs should be wide enough to allow unimpeded access, and ramps should be provided where necessary. Regulation 12(4) and paragraph 92 of this Code also deal with ramps.

162 Access between floors should not normally be by way of ladders or steep stairs. Fixed ladders or steep stairs may be used where a conventional staircase cannot be accommodated, provided they are only used by people who are capable of using them safely and any loads to be carried can be safely carried.

163 Routes should not be used by vehicles for which they are inadequate or unsuitable. Any necessary restrictions should be clearly indicated. Uneven or soft ground should be made smooth and firm if vehicles might otherwise overturn or shed their loads. Sharp or blind bends on vehicle routes should be avoided as far as possible; where they are unavoidable, measures such as one-way systems or the use of mirrors to improve vision should be considered. On vehicle routes, prominent warning should be given of any limited headroom, both in advance and at the obstruction itself. Any potentially dangerous obstructions such as overhead electric cables or pipes containing, for example, flammable or hazardous chemicals should be shielded. Screens should be provided where necessary to protect people who have to work at a place where they would be at risk from exhaust fumes, or to protect people from materials which are likely to fall from vehicles.

164 Sensible speed limits should be set and clearly displayed on vehicle routes except those used only by slow vehicles. Where necessary, suitable speed retarders such as road humps should be provided. These should always be preceded by a warning sign or a mark on the road. Arrangements should be made where necessary to avoid fork lift trucks having to pass over road humps unless the truck is of a type which can negotiate them safely.

165 Traffic routes used by vehicles should be wide enough to allow vehicles to pass oncoming or parked vehicles without leaving the route. One-way systems or restrictions on parking should be introduced as necessary. On traffic routes in existence before 1 January 1993, where it is not practical to make the route wide enough, passing places or traffic management systems should be provided as necessary.

166 Traffic routes used by vehicles should not pass close to any edge, or to anything that is likely to collapse or be left in a dangerous state if hit (such as hollow cast-iron columns and storage racking), unless the edge or thing is fenced or adequately protected.

167 The need for vehicles with poor rear visibility to reverse should be eliminated as far as possible, for example by the use of one-way systems.

168 Where large vehicles have to reverse, measures for reducing risks to pedestrians and any people in wheelchairs should be considered, such as:

(a) restricting reversing to places where it can be carried out safely;

(b) keeping people on foot or in wheelchairs away;

(c) providing suitable high visibility clothing for people who are permitted in the area;

(d) fitting reversing alarms to alert, or with a detection device to warn the driver of an obstruction or apply the brakes automatically; and

(e) employing banksmen to supervise the safe movement of vehicles.

Whatever measures are adopted, a safe system of work should operate at all times. Account should be taken of people with impaired sight or hearing.

169 If crowds of people are likely to overflow on to roadways, for example at the end of a shift, consideration should be given to stopping vehicles from using the routes at such times.

170 Where a load has to be tipped into a hopper, waste pit, or similar place, and the vehicle is liable to fall into it, substantial barriers or portable wheel stops should be provided at the end of the traffic route to prevent this type of occurrence.

Separation of people and vehicles

171 Any traffic route which is used by both pedestrians and vehicles should be wide enough to enable any vehicle likely to use the route to pass pedestrians safely. On traffic routes in existence before 1 January 1993, where it is not practical to make the route wide enough, passing places or traffic management systems should be provided as necessary. In buildings, lines should be drawn on the floor to indicate routes followed by vehicles such as fork lift trucks.

172 On routes used by automatic, driverless vehicles which are also used by pedestrians, steps should be taken to ensure that pedestrians do not become trapped by vehicles. The vehicles should be fitted with safeguards to minimise the risk of injury, sufficient clearance should be provided between the vehicles and pedestrians, and care should be taken that fixtures along the route do not create trapping hazards.

173 In doorways, gateways, tunnels, bridges, or other enclosed routes, vehicles should be separated from pedestrians by a kerb or barrier. Where necessary, for safety, separate routes through should be provided and pedestrians should be guided to use the correct route by clear marking. Such routes should be kept unobstructed. Similar measures should be taken where the speed or volume of vehicles would put pedestrians at risk.

174 Workstations should be adequately separated or shielded from vehicles.

Crossings

175 Where pedestrian and vehicle routes cross, appropriate crossing points should be provided and used. Where necessary, barriers or rails should be provided to prevent pedestrians crossing at particularly dangerous points and to guide them to designated crossing places. At crossing places where volumes of traffic are particularly heavy, the provision of suitable bridges or subways should be considered.

176 At crossing points there should be adequate visibility and open space for the pedestrian where the pedestrian route joins the vehicle route. For example, where an enclosed pedestrian route, or a doorway or staircase, joins a vehicle route there should be an open space of at least one metre from which pedestrians can see along the vehicle route in both directions (or in the case of a one-way route, in the direction of on-coming traffic). Where such a space cannot be provided barriers or rails should be provided to prevent pedestrians walking directly onto the vehicular route.

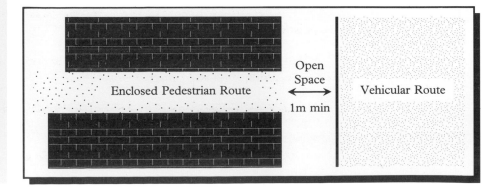

Loading bays

177 Loading bays should be provided with at least one exit point from the lower level. Wide loading bays should be provided with at least two exit points, one being at each end. Alternatively, a refuge should be provided which can be used to avoid being struck or crushed by a vehicle.

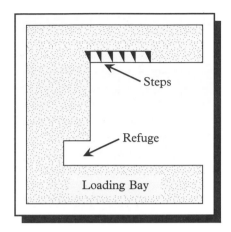

Signs

178 Potential hazards on traffic routes used by vehicles should be indicated by suitable warning signs. Such hazards may include: sharp bends, junctions, crossings, blind corners, steep gradients or roadworks.

179 Suitable road markings and signs should also be used to alert drivers to any restrictions which apply to the use of a traffic route. Adequate directions should also be provided to relevant parts of a workplace. Buildings, departments, entrances, etc should be clearly marked, where necessary, so that unplanned manoeuvres are avoided.

180 Any signs used in connection with traffic should comply with the Traffic Signs Regulations and General Directions 1981 (SI 1981 No 859) and the Highway Code for use on the public highway.

181 Further guidance on workplace transport is given in separate HSE publications[67-71].

182 There are also separate Regulations on dock work which have requirements on traffic routes[37-38].

Doors and gates

(1)   Doors and gates shall be suitably constructed (including being fitted with any necessary safety devices).

(2)   Without prejudice to the generality of paragraph (1), doors and gates shall not comply with that paragraph unless -

(a)   any sliding door or gate has a device to prevent it coming off its track during use;

(b)   any upward opening door or gate has a device to prevent it falling back;

(c)   *any powered door or gate has suitable and effective features to prevent it causing injury by trapping any person;*

(d)   *where necessary for reasons of health or safety, any powered door or gate can be operated manually unless it opens automatically if the power fails; and*

(e)   *any door or gate which is capable of opening by being pushed from either side is of such a construction as to provide, when closed, a clear view of the space close to both sides.*

183   Doors and gates which swing in both directions should have a transparent panel except if they are low enough to see over. Conventionally hinged doors on main traffic routes should also be fitted with such panels. Panels should be positioned to enable a person in a wheelchair to be seen from the other side.

184   Sliding doors should have a stop or other effective means to prevent the door coming off the end of the track. They should also have a retaining rail to prevent the door falling should the suspension system fail or the rollers leave the track.

185   Upward opening doors should be fitted with an effective device such as a counter balance or ratchet mechanism to prevent them falling back in a manner likely to cause injury.

186   Power operated doors and gates should have safety features to prevent people being injured as a result of being struck or trapped. Safety features include:

(a)   a sensitive edge, or other suitable detector, and associated trip device to stop, or reverse, the motion of the door or gate when obstructed;

(b)   a device to limit the closing force so that it is insufficient to cause injury;

(c)   an operating control which must be held in position during the whole of the closing motion. This will only be suitable where the risk of injury is low and the speed of closure is slow. Such a control, when released, should cause the door to stop or reopen immediately and should be positioned so that the operator has a clear view of the door throughout its movement.

187   Where necessary, power operated doors and gates should have a readily identifiable and accessible control switch or device so that they can be stopped quickly in an emergency. Normal on/off controls may be sufficient.

188   It should be possible to open a power operated door or gate if the power supply fails, unless it opens automatically in such circumstances, or there is an alternative way through. This does not apply to lift doors and other doors and gates which are there to prevent falls or access to areas of potential danger.

189   Where tools are necessary for manual opening they should be readily available at all times. If the power supply is restored while the door is being opened manually, the person opening it should not be put at risk.

190   The fire resistance of doors is dealt with in Building Regulations and in fire precautions legislation. Advice can be obtained from local authorities and fire authorities.

Escalators and moving walkways

(1)   Escalators and moving walkways shall:-

(a)   function safely;

(b)   be equipped with any necessary safety devices;

(c)   be fitted with one or more emergency stop controls which are easily identifiable and readily accessible.

191   There are HSE publications on the safe use and periodic thorough examination of escalators[4-5,72]. There is also a relevant British Standard[73].

Sanitary conveniences

(1)   Suitable and sufficient sanitary conveniences shall be provided at readily accessible places.

(2)   Without prejudice to the generality of paragraph (1), sanitary conveniences shall not be suitable unless -

(a)   the rooms containing them are adequately ventilated and lit;

(b)   they and the rooms containing them are kept in a clean and orderly condition; and

(c)   separate rooms containing conveniences are provided for men and women except where and so far as each convenience is in a separate room the door of which is capable of being secured from inside.

(3)   It shall be sufficient compliance with the requirement in paragraph (1) to provide sufficient sanitary conveniences in a workplace which is not a new workplace, a modification, an extension or a conversion and which, immediately before this regulation came into force in respect of it, was subject to the provisions of the Factories Act 1961, if sanitary conveniences are provided in accordance with the provisions of Part II of Schedule 1.

Washing facilities

(1)   Suitable and sufficient washing facilities, including showers if required by the nature of the work or for health reasons, shall be provided at readily accessible places.

(2)   Without prejudice to the generality of paragraph (1), washing facilities shall not be suitable unless -

(a)   they are provided in the immediate vicinity of every sanitary convenience, whether or not provided elsewhere as well;

(b)   they are provided in the vicinity of any changing rooms required by these regulations, whether or not provided elsewhere as well;

(c)   they include a supply of clean hot and cold, or warm, water (which shall be running water so far as is practicable);

(d)   they include soap or other suitable means of cleaning;

(e)   they include towels or other suitable means of drying;

(f)   the rooms containing them are sufficiently ventilated and lit;

(g)   they and the rooms containing them are kept in a clean and orderly condition; and

(h)   separate facilities are provided for men and women, except where and so far as they are provided in a room the door of which is capable of being secured from inside and the facilities in each such room are intended to be used by only one person at a time.

(3)   Paragraph (2)(h) shall not apply to facilities which are provided for washing hands, forearms and face only.

192   In paragraphs 193 - 211 'facilities' means sanitary and washing facilities, 'sanitary accommodation' means a room containing one or more sanitary conveniences and 'washing station' means a wash-basin or a section of a trough or fountain sufficient for one person.

193   Sufficient facilities should be provided to enable everyone at work to use them without undue delay. Minimum numbers of facilities are given in paragraphs 201 - 205 but more may be necessary if, for example, breaks are taken at set times or workers finish work together and need to wash before leaving.

194   Special provision should be made if necessary for any worker with a disability to have access to facilities which are suitable for his or her use.

195   The facilities do not have to be within the workplace, but they should if possible be within the building. Where arrangements are made for the use of facilities provided by someone else, for example the owner of the building, the facilities should still meet the provisions of this Code and they should be available at all material times. The use of public facilities is only acceptable as a last resort, where no other arrangement is possible.

196   Facilities should provide adequate protection from the weather.

197   Water closets should be connected to a suitable drainage system and be provided with an effective means for flushing with water. Toilet paper in a holder or dispenser and a coat hook should be provided. In the case of water closets used by women, suitable means should be provided for the disposal of sanitary dressings.

198   Washing stations should have running hot and cold, or warm water, and be large enough to enable effective washing of face, hands and forearms. Showers or baths should also be provided where the work is:

(a)   particularly strenuous;

(b)   dirty; or

(c)   results in contamination of the skin by harmful or offensive materials.

This includes, for example, work with molten metal in foundries and the manufacture of oil cake.

199 Showers which are fed by both hot and cold water should be fitted with a device such as a thermostatic mixer valve to prevent users being scalded.

200 The facilities should be arranged to ensure adequate privacy for the user. In particular:

(a) each water closet should be situated in a separate room or cubicle, with a door which can be secured from the inside;

(b) it should not be possible to see urinals, or into communal shower or bathing areas, from outside the facilities when any entrance or exit door opens;

(c) windows to sanitary accommodation, shower or bathrooms should be obscured by means of frosted glass, blinds or curtains unless it is not possible to see into them from outside; and

(d) the facilities should be fitted with doors at entrances and exits unless other measures are taken to ensure an equivalent degree of privacy.

Minimum numbers of facilities

201 Table 1 shows the minimum number of sanitary conveniences and washing stations which should be provided. The number of people at work shown in column 1 refers to the maximum number likely to be in the workplace at any one time. Where separate sanitary accommodation is provided for a group of workers, for example men, women, office workers or manual workers, a separate calculation should be made for each group.

Table 1

1 Number of people at work	2 Number of water closets	3 Number of washstations
1 to 5	1	1
6 to 25	2	2
26 to 50	3	3
51 to 75	4	4
76 to 100	5	5

202 In the case of sanitary accommodation used only by men, Table 2 may be followed if desired, as an alternative to column 2 of Table 1. A urinal may either be an individual urinal or a section of urinal space which is at least 600 mm long.

Table 2

1 Number of men at work	2 Number of water closets	3 Number of urinals
1 to 15	1	1
16 to 30	2	1
31 to 45	2	2
46 to 60	3	2
61 to 75	3	3
76 to 90	4	3
91 to 100	4	4

203 An additional water closet, and one additional washing station, should be provided for every 25 people above 100 (or fraction of 25). In the case of water closets used only by men, an additional water closet for every 50 men (or fraction of 50) above 100 is sufficient provided at least an equal number of additional urinals are provided.

204 Where work activities result in heavy soiling of face, hands and forearms, the number of washing stations should be increased to one for every 10 people at work (or fraction of 10) up to 50 people; and one extra for every additional 20 people (or fraction of 20).

205 Where facilities provided for workers are also used by members of the public the number of conveniences and washing stations specified above should be increased as necessary to ensure that workers can use the facilities without undue delay.

Remote workplaces and temporary work sites

206 In the case of remote workplaces without running water or a nearby sewer, sufficient water in containers for washing, or other means of maintaining personal hygiene, and sufficient chemical closets should be provided. Chemical closets which have to be emptied manually should be avoided as far as possible. If they have to be used a suitable deodorising agent should be provided and they should be emptied and recharged at suitable intervals.

207 In the case of temporary work sites, which are referred to in paragraph 12 of this document, regulation 3(2) requires that suitable and sufficient sanitary conveniences and washing facilities should be provided so far as is reasonably practicable. As far as possible, water closets and washing stations which satisfy this Code should be provided. In other cases, mobile facilities should be provided wherever possible. These should if possible include flushing sanitary conveniences and running water for washing and meet the other requirements of this Code.

Ventilation, cleanliness and lighting

208 Any room containing a sanitary convenience should be well ventilated, so that offensive odours do not linger. Measures should also be taken to prevent odours entering other rooms. This may be achieved by, for example, providing a ventilated area between the room containing the convenience and the other room. Alternatively it may be possible to achieve it by mechanical ventilation or, if the room containing the convenience is well sealed from the workroom and has a door with an automatic closer, by good natural ventilation. However, no room containing a sanitary convenience should communicate directly with a room where food is processed, prepared or eaten.

209 Arrangements should be made to ensure that rooms containing sanitary conveniences or washing facilities are kept clean. The frequency and thoroughness of cleaning should be adequate for this purpose. The surfaces of the internal walls and floors of the facilities should normally have a surface which permits wet cleaning, for example ceramic tiling or a plastic coated surface. The rooms should be well lit; this will also facilitate cleaning to the necessary standard and give workers confidence in the cleanliness of the facilities. Responsibility for cleaning should be clearly established, particularly where facilities are shared by more than one workplace.

Other Regulations and publications

210 Legionnaires' disease is caused by bacteria which may be found where water stands for long periods at lukewarm or warm temperatures in, for example, tanks or little used pipes. Separate HSE publications are available[28-29].

211 Other Regulations and Approved Codes of Practice on the control of substances hazardous to health also deal with washing facilities[15-22]. Information about the requirements of food hygiene legislation can be obtained from the Environmental Health Department of local authorities.

Drinking water

(1) An adequate supply of wholesome drinking water shall be provided for all persons at work in the workplace.

(2) Every supply of drinking water required by paragraph (1) shall –

(a) be readily accessible at suitable places; and

(b) be conspicuously marked by an appropriate sign where necessary for reasons of health or safety.

(3) Where a supply of drinking water is required by paragraph (1), there shall also be provided a sufficient number of suitable cups or other drinking vessels unless the supply of drinking water is in a jet from which persons can drink easily.

212 Drinking water should normally be obtained from a public or private water supply by means of a tap on a pipe connected directly to the water main. Alternatively, drinking water may be derived from a tap on a pipe connected directly to a storage cistern which complies with the requirements of the UK Water Bye-laws. In particular, any cistern, tank or vessel used as a supply should be well covered, kept clean and tested and disinfected as necessary. Water should only be provided in refillable containers where it cannot be obtained directly from a mains supply. Such containers should be suitably enclosed to prevent contamination and should be refilled at least daily.

213 Drinking water taps should not be installed in places where contamination is likely, for example in a workshop where lead is handled or processed. As far as is reasonably practicable they should also not be installed in sanitary accommodation.

214 Drinking cups or beakers should be provided unless the supply is by means of a drinking fountain. In the case of non-disposable cups a facility for washing them should be provided nearby.

215 Drinking water supplies should be marked as such if people may otherwise drink from supplies which are not meant for drinking. Marking is not necessary if non-drinkable cold water supplies are clearly marked as such.

216 Any cold water supplies which are likely to be grossly contaminated, as in the case of supplies meant for process use only, should be clearly marked by a suitable sign.

Accommodation for clothing

(1)   Suitable and sufficient accommodation shall be provided –

(a)   for the clothing of any person at work which is not worn during working hours; and

(b)   for special clothing which is worn by any person at work but which is not taken home.

(2)   Without prejudice to the generality of paragraph (1), the accommodation mentioned in that paragraph shall not be suitable unless –

(a)   where facilities to change clothing are required by regulation 24, it provides suitable security for the clothing mentioned in paragraph (1)(a);

(b)   where necessary to avoid risks to health or damage to the clothing, it includes separate accommodation for clothing worn at work and for other clothing;

(c)   so far as is reasonably practicable, it allows or includes facilities for drying clothing; and

(d)   it is in a suitable location.

217   Special work clothing includes all clothing which is only worn at work such as overalls, uniforms, thermal clothing and hats worn for food hygiene purposes.

218   Accommodation for work clothing and workers' own personal clothing should enable it to hang in a clean, warm, dry, well-ventilated place where it can dry out during the course of a working day if necessary. If the workroom is unsuitable for this purpose then accommodation should be provided in another convenient place. The accommodation should consist of, as a minimum, a separate hook or peg for each worker.

219   Where facilities to change clothing are required by regulation 24, effective measures should be taken to ensure the security of clothing. This may be achieved, for example, by providing a lockable locker for each worker.

220   Where work clothing (including personal protective equipment) which is not taken home becomes dirty, damp or contaminated due to the work it should be accommodated separately from the worker's own clothing. Where work clothing becomes wet, the facilities should enable it to be dried by the beginning of the following work period unless other dry clothing is provided.

221   Separate Regulations deal with personal protective equipment at work in greater detail[13-14].

222   Other Regulations and Approved Codes of Practice on the control of substances hazardous to health also deal with accommodation for clothing[15-22]. Information about the requirements for food hygiene legislation can be obtained from the Environmental Health Department of local authorities.

Facilities for changing clothing

(1)   Suitable and sufficient facilities shall be provided for any person at work in the workplace to change clothing in all cases where-

(a)   the person has to wear special clothing for the purpose of work; and

(b)   the person can not, for reasons of health or propriety, be expected to change in another room.

(2)   Without prejudice to the generality of paragraph (1), the facilities mentioned in that paragraph shall not be suitable unless they include separate facilities for, or separate use of facilities by, men and women where necessary for reasons of propriety.

223   A changing room or rooms should be provided for workers who change into special work clothing (see paragraph 217) and where they remove more than outer clothing. Changing rooms should also be provided where necessary to prevent workers' own clothing being contaminated by a harmful substance.

224   Changing facilities should be readily accessible from workrooms and eating facilities, if provided. They should be provided with adequate seating and should contain, or communicate directly with, clothing accommodation and showers or baths if provided. They should be constructed and arranged to ensure the privacy of the user.

225   The facilities should be large enough to enable the maximum number of persons at work expected to use them at any one time, to do so without overcrowding or unreasonable delay. Account should be taken of starting and finishing times of work and the time available to use the facilities.

226   Other Regulations and Approved Codes of Practice on the control of substances hazardous to health also deal with changing facilities [15-22].

Facilities for rest and to eat meals

(1)   Suitable and sufficient rest facilities shall be provided at readily accessible places.

(2)   Rest facilities provided by virtue of paragraph (1) shall –

(a)   where necessary for reasons of health or safety include, in the case of a new workplace, an extension or a conversion, rest facilities provided in one or more rest rooms, or, in other cases, in rest rooms or rest areas;

(b)   include suitable facilities to eat meals where food eaten in the workplace would otherwise be likely to become contaminated.

(3)   Rest rooms and rest areas shall include suitable arrangements to protect non-smokers from discomfort caused by tobacco smoke.

(4)   Suitable facilities shall be provided for any person at work who is a pregnant woman or nursing mother to rest.

(5)   Suitable and sufficient facilities shall be provided for persons at work to eat meals where meals are regularly eaten in the workplace.

227 For workers who have to stand to carry out their work, suitable seats should be provided for their use if the type of work gives them an opportunity to sit from time to time.

228 Suitable seats should be provided for workers to use during breaks. These should be in a suitable place where personal protective equipment (for example respirators or hearing protection) need not be worn. In offices and other reasonably clean workplaces, work seats or other seats in the work area will be sufficient, provided workers are not subject to excessive disturbance during breaks, for example, by contact with the public. In other cases one or more separate rest areas should be provided (which in the case of new workplaces, extensions and conversions should include a separate rest room).

229 Rest areas or rooms provided in accordance with regulation 25(2) should be large enough, and have sufficient seats with backrests and tables, for the number of workers likely to use them at any one time.

230 If workers frequently have to leave their work area, and to wait until they can return, there should be a suitable rest area where they can wait.

231 Where workers regularly eat meals at work suitable and sufficient facilities should be provided for the purpose. Such facilities should also be provided where food would otherwise be likely to be contaminated, including by dust or water, for example:

(a) cement works, clay works, foundries, potteries, tanneries, and laundries;

(b) the manufacture of glass bottles and pressed glass articles, sugar, oil cake, jute, and tin or terne plates; and

(c) glass bevelling, fruit preserving, gut scraping, tripe dressing, herring curing, and the cleaning and repairing of sacks.

232 Seats in work areas can be counted as eating facilities provided they are in a sufficiently clean place and there is a suitable surface on which to place food. Eating facilities should include a facility for preparing or obtaining a hot drink, such as an electric kettle, a vending machine or a canteen. Workers who work during hours or at places where hot food cannot be obtained in, or reasonably near to, the workplace should be provided with the means for heating their own food.

233 Eating facilities should be kept clean to a suitable hygiene standard. Responsibility for cleaning should be clearly allocated. Steps should be taken where necessary to ensure that the facilities do not become contaminated by substances brought in on footwear or clothing. If necessary, adequate washing and changing facilities should be provided in a conveniently accessible place.

234 Canteens or restaurants may be used as rest facilities, provided that there is no obligation to purchase food in order to use them.

235 Good hygiene standards should be maintained in those parts of rest facilities used for eating or preparing food and drinks.

236 The subject of eating in the workplace is also dealt with in other Regulations concerning asbestos, lead, and ionising radiations, and in Approved Codes of Practice on the control of substances hazardous to health and on work in potteries[15-22,74].

Facilities for pregnant women and nursing mothers

237   Facilities for pregnant women and nursing mothers to rest should be conveniently situated in relation to sanitary facilities and, where necessary, include the facility to lie down.

238   There is an HSE guidance sheet on health aspects of pregnancy[75].

Prevention of discomfort caused by tobacco smoke

239   Rest areas and rest rooms should be arranged to enable employees to use them without experiencing discomfort from tobacco smoke. Methods of achieving this include:

(a)   the provision of separate areas or rooms for smokers and non-smokers; or

(b)   the prohibition of smoking in rest areas and rest rooms.

240   Passive smoking in the workplace is dealt with in a separate HSE publication[76].

Exemption certificates

(1)   The Secretary of State for Defence may, in the interests of national security, by a certificate in writing exempt any of the home forces, any visiting force or any headquarters from the requirements of these Regulations and any exemption may be granted subject to conditions and to a limit of time and may be revoked by the said Secretary of State by a further certificate in writing at any time.

(2)   In this Regulation -

(a)   "the home forces" has the same meaning as in section 12(1) of the Visiting Forces Act 1952;

(b)   "headquarters" has the same meaning as in article 3(2) of the Visiting Forces and International Headquarters (Application of Law) Order 1965;

(c)   "visiting force" has the same meaning as it does for the purposes of any provision of Part I of the Visiting Forces Act 1952.

Repeals, saving and revocations

(1)   The enactments mentioned in column 2 of Part I of Schedule 2 are repealed to the extent specified in column 3 of that Part.

(2)   Nothing in this regulation shall affect the operation of any provision of the Offices, Shops and Railway Premises Act 1963 as that provision has effect by virtue of section 90(4) of that Act.

(3)   The instruments mentioned in column 1 of Part II of Schedule 2 are revoked to the extent specified in column 3 of that Part.

Regulations 10 and 20

Provisions applicable to factories which are not new workplaces, extensions or conversions

Part I - Space

1   No room in the workplace shall be so overcrowded as to cause risk to the health or safety of persons at work in it.

2   Without prejudice to the generality of paragraph 1, the number of persons employed at a time in any workroom shall not be such that the amount of cubic space allowed for each is less than 11 cubic metres.

3   In calculating for the purposes of this Part of this Schedule the amount of cubic space in any room no space more than 4.2 metres from the floor shall be taken into account and, where a room contains a gallery, the gallery shall be treated for the purposes of this Schedule as if it were partitioned off from the remainder of the room and formed a separate room.

Part II - Number of sanitary conveniences

4   In workplaces where females work, there shall be at least one suitable water closet for use by females only for every 25 females.

5   In workplaces where males work, there shall be at least one suitable water closet for use by males only for every 25 males.

6   In calculating the number of males or females who work in any workplace for the purposes of this Part of this Schedule, any number not itself divisible by 25 without fraction or remainder shall be treated as the next number higher than it which is so divisible.

Regulation 27

Repeals and revocations

Part I - Repeals

1 *Chapter*	2 *Short title*	3 *Extent of repeal*
1961 c.34	The Factories Act 1961	Section 1 to 7, 18, 28, 29, 57 to 60 and 69
1963 c.41	The Offices, Shops and Railway Premises Act 1963	Sections 4 to 16
1956 c.49	The Agriculture (Safety, Health and Welfare Provisions) Act 1956	Section 3 and 5 and, in section 25, sub-sections (3) and (6)

Part II - Revocations

1 *Title*	2 *Reference*	3 *Extent of revocation*
The Flax and Tow Spinning and Weaving Regulations 1906	S.R. & O. 1906/177, amended by S.I. 1988/1657	Regulations 3, 8, 10, 11 and 14
The Hemp Spinning and Weaving Regulations 1907	S.R. & O. 1907/660, amended by S.I. 1988/1657	Regulations 3 to 5 and 8
Order dated 5 October 1917 (the Tin or Terne Plates Manufacture Welfare Order 1917)	S.R. & O. 1917/1035	The whole Order
Order dated 15 May 1918 (the Glass Bottle, etc. Manufacture Welfare Order 1918)	S.R. & O. 1918/558	The whole Order
Order dated 15 August 1919 (the Fruit Preserving Order 1919)	S.R. & O. 1919/1136, amended by S.I. 1988/1657	The whole Order
Order dated 23 April 1920 (the Laundries Welfare Order 1920)	S.R. & O. 1920/654	The whole Order
Order dated 28 July 1920 (the Gut Scraping, Tripe Dressing, etc. Welfare Order 1920)	S.R. & O. 1920/1437	The whole Order
Order dated 9 September 1920 (the Herring Curing (Norfolk and Suffolk) Welfare Order 1920)	S.R. & O. 1920/1662	The whole Order
Order dated 3 March 1921 (the Glass Bevelling Welfare Order 1921)	S.R. & O. 1921/288	The whole Order
The Herring Curing (Scotland) Welfare Order 1926	S.R. & O. 1926/535 (S.24)	The whole Order

Part II - Revocations (continued)

1 Title	2 Reference	3 Extent of revocation
The Herring Curing Welfare Order 1927	S.R. & O. 1927/813, amended by S.I. 1960/1690 and 917	The whole Order
The Sacks (Cleaning and Repairing) Welfare Order 1927	S.R. & O. 1927/860	The whole Order
The Horizontal Milling Machines Regulations 1928	S.R. & O. 1928/548	The whole Regulations
The Cotton Cloth Factories Regulations 1929	S.I. 1929/300	Regulations 5 to 10, 11 and 12
The Oil Cake Welfare Order 1929	S.R. & O. 1929/534	Articles 3 to 6
The Cement Works Welfare Order 1930	S.R. & O. 1930/94	The whole Order
The Tanning Welfare Order 1930	S.R. & O. 1930/312	The whole Order
The Kiers Regulations 1938	S.R.& O. 1938/106, amended by S.I. 1981/1152	Regulations 12 to 15
The Sanitary Accommodation Regulations 1938	S.R. & O. 1938/611, amended by S.I.1974/426	The whole Regulations
The Clay Works (Welfare) Special Regulations 1948	S.I. 1948/1547	Regulations 3, 4, 6, 8 and 9
The Jute (Safety, Health and Welfare) Regulations 1948	S.I.1948/1696, amended by S.I.1988/1657	Regulations 11, 13, 14 to 16 and 19 to 26
The Pottery (Health and Welfare) Special Regulations 1950	S.I.1950/65, amended by S.I.1963/879, 1973/36, 1980/1248, 1982/877, 1988/1657, 1989/2311 and 1990/305	Regulation 15
The Iron and Steel Foundries Regulations 1953	S.I.1953/1464, amended by S.I.1974/1681 and 1981/1332	The whole Regulations
The Washing Facilities (Running Water) Exemption Regulations 1960	S.I.1960/1029	The whole Regulations
The Washing Facilities (Miscellaneous Industries) Regulations 1960	S.I.1960/1214	The whole Regulations
The Factories (Cleanliness of Walls and Ceilings) Order 1960	S.I.1960/1794, amended by S.I.1974/427	The whole Order
The Non-ferrous Metals (Melting and Founding) Regulations 1962	S.I.1962/1667, amended by S.I.1974/1681, 1981/1332 and 1988/165	Regulations 5, 6 to 10, 14 to 17 and 20
The Offices, Shops and Railway Premises Act 1963 (Exemption No 1) Order 1964	S.I.1964/964	The whole Order
The Washing Facilities Regulations 1964	S.I.1964/965	The whole Regulations

Part II - Revocations (continued)

1 Title	2 Reference	3 Extent of revocation
The Sanitary Conveniences Regulations 1964	S.I.1964/966, amended by S.I.1982/827	The whole Regulations
The Offices, Shops and Railway Premises Act 1963 (Exemption No 7) Order 1968	S.I.1968/1947, amended by S.I. 1982/827	The whole Order
The Abrasive Wheels Regulations 1970	S.I.1970/535	Regulation 17
The Sanitary Accommodation (Amendment) Regulations 1974	S.I.1974/426	The whole Regulations
The Factories (Cleanliness of Walls and Ceilings) (Amendment) Regulations 1974	S.I.1974/427	The whole Regulations
The Woodworking Machines Regulations 1974	S.I.1974/903, amended by S.I.1978/1126	Regulations 10 to 12
The Offices, Shops and Railway Premises Act 1963 etc. (Metrication) Regulations 1982	S.I.1982/827	The whole Regulations

References

Where reference is made to a British Standard, there may also be an equivalent European Standard.

1   BS 5810:1979 *Code of practice for access for the disabled to buildings*

2   *Management of Health and Safety at Work Regulations* 1992 SI 1992 No 2051 HMSO ISBN 0 11 025051 6

3   HSE *Management of health and safety at work: Approved code of practice* L21 HMSO 1992 ISBN 0 11 886330 4

4   HSE *Safety in the use of escalators* PM 34 HMSO 1983 ISBN 0 11 883572 6

5   HSE *Escalators: periodic thorough examination* PM 45 HMSO 1984 ISBN 0 11 883595 5

6   HSE *Prevention of falls to window cleaners* GS 25 HMSO 1992 ISBN 0 11 885682 0

7   HSE *Suspended access equipment* PM 30 HMSO 1983 ISBN 0 11 883577 7

8   BS 8210:1986 *Guide to building maintenance management*

9   Chartered Institution of Building Services Engineers *Maintenance management for building services* TM 17

10   Chartered Institution of Building Services Engineers *Operating and maintenance manuals for building services installations* BAG1 1987

11   *Provision and Use of Work Equipment Regulations 1992* SI 1992 No 2932 ISBN 0 11 025849 5

12   HSE *Work equipment.* Guidance on the Provision and Use of Work Equipment Regulations 1992 L22 HMSO 1992 ISBN 0 11 886332 0

13   *Personal Protective Equipment at Work Regulations 1992* SI 1992 No 2966 ISBN 0 11 025832 0

14   HSE *Personal protective equipment at work.* Guidance on the Personal Protective Equipment at Work Regulations 1992 L25 HMSO 1992 ISBN 0 11 886334 7

15   *The Control of Substances Hazardous to Health Regulations 1988* SI 1988/1657 HMSO ISBN 0 11 087657 1

16   HSE *Control of substances hazardous to health and control of carcinogenic substances Control of Substances Hazardous to Health (COSHH) Regulations 1988.* Approved codes of practice L5 (3rd ed) HMSO 1992 ISBN 0 11 885698 7

17   *The Control of Asbestos at Work Regulations* 1987 SI 1987/2115 HMSO ISBN 0 11 078115 5

18 HSE *Work with asbestos insulation, asbestos coating and asbestos insulating board.* Revised approved code of practice COP 3 HMSO 1988 ISBN 0 11 883979 9
Control of asbestos at work: The Control of Asbestos at Work Regulations 1987 COP 21 HMSO 1987 ISBN 0 11 883984 5

19 *The Control of Lead at Work Regulations 1980* SI 1980/1248 HMSO ISBN 0 11 007248 0

20 HSE *Control of lead at work approved code of practice.* Revised June 1985 COP 2 HMSO 1985 ISBN 0 11 883780 X

21 *Ionising Radiation Regulations 1985* SI 1985/1333 HMSO ISBN 0 11 057333 1

22 HSE *Protection of persons against ionising radiation arising from any work activity: the Ionising Radiations Regulations 1985: approved code of practice* COP 17 1985 HMSO ISBN 0 11 883838 5

23 HSE *Ventilation of the workplace* EH 22(Rev) HMSO 1988 ISBN 0 11 885403 8

24 HSE *Measurement of air change rates in factories and offices* MDHS 73 HMSO 1992 ISBN 0 11 885693 6

25 Chartered Institution of Building Services Engineers Design CIBSE guide: Volume A *Design data* GVA 1986

26 Chartered Institution of Building Services Engineers *Air filtration and natural ventilation* GS A4 1988 ISBN 0 90 095329 2

27 Chartered Institution of Building Services Engineers *Ventilation effectiveness in mechanical ventilation systems* BTN1 1988 ISBN 0 86 022189 X

28 HSE *Approved Code of Practice for the prevention or control of legionellosis (including legionnaires' disease)* L8 HMSO 1991 ISBN 0 11 885659 6)

29 HSE *The control of legionellosis including legionnaires' disease* HS(G)70 HMSO 1991 ISBN 0 11 885660 X

30 Chartered Institution of Building Services Engineers *Minimising the risk of legionnaires' disease* TM 13 1991 ISBN 0 90 095352 7

31 HSE *Lighting at work* HS(G)38 HMSO 1987 ISBN 0 11 883964 0

32 Chartered Institution of Building Services Engineers *Code for interior lighting* CIL 1984 ISBN 0 90 095327 6 (New edition due 1993)

33 Chartered Institution of Building Services Engineers *Lighting guide: The industrial environment* LG1 1989 ISBN 0 90 095338 1

34 Chartered Institution of Building Services Engineers *Hospitals and health care buildings* 1989 ISBN 0 90 095341 1

35 Chartered Institution of Building Services Engineers *Lighting guide: Areas for visual display terminals* LG3 1989 ISBN 0 90 095341 1

36 Chartered Institution of Building Services Engineers *Lighting in hostile and hazardous environments* AG HHE 1983 ISBN 0 90 095326 8

37 *Docks Regulations 1988* SI 1988/1655 HMSO ISBN 0 11 087655 5

38 HSE *Safety in docks: Docks Regulations 1988: Approved code of practice with regulations and guidance* COP 25 HMSO 1988
ISBN 0 11 885456 9

39 *Health and Safety (Display Screen Equipment) Regulations 1992*
SI 1992 No 2792 ISBN 0 11 025919 X

40 HSE *Display screen equipment work.* Guidance on the Health and Safety
(Display Screen Equipment) Regulations 1992 L26 HMSO 1992
ISBN 0 11 886331 2

41 HSE *Visual display units* HMSO 1983 ISBN 0 11 883685 4

42 HSE *Working with VDUs* IND(G)36L 1992*

43 *Electricity at Work Regulations 1989* No 635 HMSO
ISBN 0 11 096635 X

44 HSE *Memorandum of guidance on the Electricity at Work Regulations 1989*
HS(R)25 HMSO 1989 ISBN 0 11 883963 2

45 HSE *Seating at work* HS(G) 57 HMSO 1991 ISBN 0 11 885431 3

46 HSE *Ergonomics at work* IND(G)90L 1990*

47 HSE *Human factors in industrial safety* HS(G) 48 HMSO 1989
ISBN 0 11 885486 0

48 HSE *Work related upper limb disorders: a guide to prevention*
HS(G) 60 HMSO 1990 ISBN 0 11 885565 4

49 HSE *Watch your step: prevention of slipping, tripping and falling accidents at work* HMSO 1985 ISBN 0 11 883782 6

50 HSE *Storage of approved pesticides: guidance for farmers and other professional users* CS 19 HMSO 1988 0 11 885406 2

51 BS 6399:Part 1:1984 *Design loading for buildings: code of practice for dead and imposed loads*

52 HSE *Effluent storage on farms* GS 12 HMSO 1981
ISBN 0 11 883386 3

53 BS 6180:1982 *Code of practice for protective barriers in and about buildings*

54 BS 5395:1985 *Code of practice for the design of industrial type stairs, permanent ladders and walkways*

55 BS 4211:1987 *Specification for ladders for permanent access to chimneys, other high structures, silos and bins*

56 HSE *Safety in roofwork* HS(G)33 HMSO 1987 ISBN 0 11 8883922 5

57   BS 6399:Part 3:1988 *Design loading for buildings: code of practice for imposed roof loads*

58   HSE *Stacking of bales in agriculture* IND(G)125L 1992*

59   HSE *Health and safety in retail and wholesale warehouses* HS(G)76 HMSO ISBN 0 11 885731 2

60   BS 1397:1979 *Specification for industrial safety belts, harnesses and safety lanyards*

61   BS 5845:1991 *Specification for permanent anchors for industrial safety belts and harnesses*

62   *Shipbuilding and Ship-repairing Regulations 1960* SI 1960/1248 HMSO 1960

63   *Agriculture (Safeguarding of Workplaces) Regulations 1959* SI 1959/428

64   BS 6206:1981 *Specification for impact performance requirements for flat safety glass and safety plastics for use in buildings*

65   BS 6262:1982 *Code of practice for glazing in buildings*

66   BS 8213:Part 1:1991 *Windows, doors and rooflights: code of practice for safety in use and during cleaning of windows (including guidance on cleaning materials and methods)*

67   HSE *Road transport in factories and similar workplaces* GS9(Rev) HMSO 1992 ISBN 0 11 885732 0

68   HSE *Safety in working with lift trucks* HS(G)6 (Rev) HMSO 1993 ISBN 0 11 886395 9 (Due early 1993)

69   HSE *Container terminals: safe working practice* HS(G)7 HMSO 1980 ISBN 0 11 883302 2

70   HSE *Danger! Transport at work* IND(G)22L 1985*

71   HSE *Transport kills: a study of fatal accidents in industry* 1978-80 HMSO 1982 ISBN 0 11 883659 5

72   HSE *Ergonomic aspects of escalators used in retail organisations* CRR12/1989 HMSO 1989 ISBN 0 11 885938 2

73   BS 5656:1983 *Safety rules for the construction and installation of escalators and passenger conveyors*

74   HSE *Control of substances hazardous to health in the production of pottery. The Control of Substances Hazardous to Health Regulations 1988. The Control of Lead at Work Regulations 1980. Approved Code of Practice* COP 41 HMSO 1990 ISBN 0 11 885530 1

75   HSE *Occupational health aspects of pregnancy* MA 6 1989 (available free from local HSE offices)

76   HSE *Passive smoking at work* IND(G)63L* (Rev) 1992

Addresses

British and European Standards are available from British Standards Institution, Sales Department, Linford Wood, Milton Keynes, MK14 6LE.

The Chartered Institution of Building Services Engineers publications are available from Delta House, 222 Balham High Road, London, SW12 9BS.

HSE publications are available from HMSO bookshops, except those marked with an asterisk which are free from HSE Information Centre.

Extracts from relevant health and safety legislation

Health and Safety at Work etc Act 1974 - Section 2

"(1) It shall be the duty of every employer to ensure, so far as is reasonably practicable, the health, safety and welfare at work of all his employees.

(2) Without prejudice to the generality of an employer's duty under the preceding subsection, the matters to which that duty extends include in particular -

    (a) the provision and maintenance of plant and systems of work that are, so far as is reasonably practicable, safe and without risks to health;

    (b) arrangements for ensuring, so far as is reasonably practicable, safety and absence of risks to health in connection with the use, handling, storage and transport of articles and substances;

    (c) the provision of such information, instruction, training and supervision as is necessary to ensure, so far as is reasonably practicable, the health and safety at work of his employees;

    (d) so far as is reasonably practicable as regards any place of work under the employer's control, the maintenance of it in a condition that is safe and without risks to health and the provision and maintenance of means of access to and egress from it that are safe and without such risks;

    (e) the provision and maintenance of a working environment for his employees that is, so far as is reasonably practicable, safe, without risks to health, and adequate as regards facilities and arrangements for their welfare at work."

Health and Safety at Work etc Act 1974 - Section 4

"(1) This section has effect for imposing on persons duties in relation to those who -

    (a) are not their employees; but

    (b) use non-domestic premises made available to them as a place of work or as a place where they may use plant or substances provided for their use there,

and applies to premises so made available and other non-domestic premises used in connection with them.

(2) It shall be the duty of each person who has, to any extent, control of premises to which this section applies or of the means of access thereto or egress therefrom or of any plant or substance in such premises to take such measures as it is reasonable for a person in his position to take to ensure, so far as is reasonably practicable, that the premises, all means of access thereto or egress therefrom available for use by persons using the premises, and any plant or substance in the premises or, as the case may be, provided for use there, is or are safe and without risks to health.

(3)   Where a person has, by virtue of any contract or tenancy, an obligation of any extent in relation to -

   (a)   the maintenance or repair of any premises to which this section applies or any means of access thereto or egress therefrom; or

   (b)   the safety of or the absence of risks to health arising from plant or substances in any such premises;

that person shall be treated, for the purposes of subsection (2) above, as being a person who has control of the matters to which his obligations extends.

(4)   Any reference in this section to a person having control of any premises or matter is a reference to a person having control of the premises or matter in connection with the carrying on by him of a trade, business or other undertaking (whether for profit or not)."

Factories Act 1961 - Section 175(5)

"Any workplace in which, with the permission of or under agreement with the owner or occupier, two or more persons carry on any work which would constitute a factory if the persons working therein were in the employment of the owner or occupier, shall be deemed to be a factory for the purposes of this Act, and, in the case of any such workplace not being a tenement factory or part of a tenement factory, the provisions of this Act shall apply as if the owner or occupier of the workplace were the occupier of the factory and the persons working therein were persons employed in the factory."

Printed in the United Kingdom for HSE, published by HMSO